WHEREBY WE THRIVE

A HISTORY OF AMERICAN FARMING, 1607–1972

WHEREBY WE THRIVE

A HISTORY OF AMERICAN FARMING, 1607–1972

JOHN T. SCHLEBECKER

THE IOWA STATE UNIVERSITY PRESS, AMES, IOWA

1 9 7 5

JOHN T. SCHLEBECKER, Curator, Agriculture and Mining, Smithsonian Institution, received the B.A. degree from Hiram College in 1949, the master's degree from Harvard University in 1951, and the Ph.D degree from the University of Wisconsin (Madison) in 1954. He formerly held positions as assistant professor of history at Montana State University (1954–1956) and as associate professor of history at Iowa State University (1956–1965). His books include, among others, *A History of Dairy Journalism in the United States, 1819–1950* (with A. W. Hopkins); *Cattle Raising on the Plains, 1900–1961; A History of American Dairying;* and a *Bibliography of Books and Pamphlets on the History of Agriculture in the United States, 1607–1967.*

© 1975 The Iowa State University Press
Ames, Iowa 50010. All rights reserved

Composed and printed by
The Iowa State University Press

First edition, 1975
Second printing, 1976

Library of Congress Cataloging in Publication Data

Schlebecker, John T
    Whereby we thrive: a history of American farming,
1607 to 1972.

    Bibliography: p.
    Includes index.
    1. Agriculture—United States—History.   I. Title.
S441.S43   630′.973   74–19455
ISBN 0–8138–0090–0

CONTENTS ⚜

viii

PREFACE 🪶

THIS HISTORY is not a source of new interpretations or information. I have used mostly secondary material in bringing together what other historians have written about American agricultural history. Through obviously pertinent events and people, my story centers on what farmers themselves achieved and what others achieved for them.

However, no narrative history can be told without interpretation and analysis. Storytellers must at least decide what goes into their tales; this first step alone is analytic. They may even take a next interpretive step of trying to differentiate the more important from the less important events—and this I have done.

The topics are collected within large time periods: 1607–1783 (177 years); 1783–1861 (79 years); 1861–1914 (54 years); 1914–1945 (32 years); and 1945–1972 (28 years).

Since this history deals mainly with the work of ordinary people doing everyday tasks, it becomes basically social history. But American farmers consistently strove to make money, so agricultural history also developed as a form of economic history. Because technological history is subsumed under economics, the invention of any machine assumes importance primarily because of its effect on the lives of the people who used the device. In America, even subsistence farmers were regarded as putative commercial farmers. Some confusion on this matter has arisen because historians have tended to define commercial farming in terms of the commercial farmers of their own day. Obviously a 17th-century New England farmer was not a commercial farmer if compared to a 19th-century cotton planter. Considered within their own milieu, however, most of them can be regarded as commercial farmers. In short, this history of American agriculture becomes a history of commercial farming.

We should note that commercial farming is not necessarily specialized farming. A farmer can be a general farmer and still be a commercial farmer. He can raise a great variety of things and his operation can be quite commercialized. Commercial means merely that the farm produce enters into commerce.

ix

Nor, in defining commercial farming, can size of operation be decisive. A celery producer with five acres, selling all he grows, must be considered a commercial farmer. On the other hand, an 18th-century Virginian with 2,000 acres of land, but only a corn patch and three hogs, cannot because mere size of farm does not matter.

Nor can the definition rest entirely on the amount a farmer sells in any given year. A bonanza wheat farmer with 100,000 acres of wheat all destroyed by grasshoppers would by some definitions not be a commercial farmer because he had not sold anything. Furthermore, what the farmer intends to do cannot be the standard for measuring his commercial involvement. It must be defined in terms of what the farmer actually did, not what he thought he was doing or wanted to do.

Commercial farming can be subject to even more interpretations. A farmer of the 17th century might qualify even though most of his wealth came from the sale of land. Similarly, a 20th-century farmer could be a commercial farmer even though most of his income came from his insurance agency or some other nonfarm enterprise. No definition except one really suits all cases. A farmer can be called a commercial farmer if he sells any farm commodity and receives payment. Under this definition, American farmers were overwhelmingly commercial farmers from the 17th century on. Their story is outlined in this narrative.

The topical divisions permit readers to pursue only those topics that interest them. Three general topics recur in each time segment: land, markets, and technology and science, and each topic is covered for every period. Such subtopics as governmental administrative activity and the political activity of farmers are merged with the major topics as seemed appropriate.

ACKNOWLEDGMENTS

My debt to other scholars is great yet diffuse and involves too many to list. However, I must mention John C. Greene, professor of history of science, University of Connecticut, who in 1960 when he was with Iowa State University, urged me to write a survey of American agricultural history. It took me longer than either of us thought likely, but I am still grateful for the suggestion.

I also gratefully acknowledge the information and interpretations provided by the lectures, seminars, and conversations of Vernon Carstensen and Frederick Merk, my teachers.

Photographs are courtesy of the Smithsonian Institution unless otherwise noted.

HISTORY celebrates the battlefields whereon
we meet our death, but scorns to speak of
the plowed fields whereby we thrive. It
knows the names of the kings' bastards but
cannot tell us the origin of wheat. This is
the way of human folly.

J. H. FABRE

CHAPTER ONE

LAND FOR FARMING ✠ 1607–1763

THE FIRST ENGLISH SETTLERS in the New World hoped for wealth from a variety of enterprises, but they apparently expected little from agriculture except subsistence. Few experienced farmers came with the first immigrants, and the settlers counted on regular food shipments from England for the first year at least. Settlers at Jamestown, Virginia, in 1607 spent more time exploring for gold, spices, and furs to make their fortunes than in farming for survival. Seventeenth-century farmers brought few specialized skills or tools and what they did bring seldom suited their needs.

These first planters found plenty of fertile land, but in spite of its abundance it was difficult to acquire. The difficulties arose from European law and economic theory as well as Indian occupation of the land. The settlers and their backers in England had to modify both land laws and theory before a truly flourishing agriculture took shape.

To further its colonial interests, the English government granted monopoly rights to certain companies for trade and settlement in various parts of the world. The government felt it could more easily control one monopolistic company than a variety of small firms or a mass of people. In Virginia, the Crown granted a royal monopoly to the Virginia Company of London in 1606. This company owned the land, and most of the first settlers came as servants of the company, indentured for seven years. As their indentures ended, beginning in 1615, the company allowed them to become share-cropping tenants on company land. Indentured servants of the company continued to come to Virginia until the Crown took over the colony in 1624.

Increasingly larger numbers arrived on company land as independent settlers. Many of these received large private holdings of land called *hundreds* located on company territory. To encourage settlers to leave England and to regularize the freeing of indentured servants, the company devised the headright system of land disposal

3

in 1619. Each farmer could have 50 acres for himself and 50 acres for each person he brought to America. Giving land to those who came at their own expense encouraged immigration. The system effectively distributed the land and encouraged the dispersal of the settlers. In 1624, after years of wrangling, financial losses, and successful colonization, the Crown replaced the company in Virginia; but the headright system continued.

The headright system began in Virginia and was adopted by colonies of the South and Middle Atlantic. Clever men soon discovered how to expand on the rewards. Three headrights could be collected on one immigrant: the ship captain could collect on every indentured servant and on the members of his crew; the dealer in indentured servants could collect; and the final purchaser of the indentured servant could collect. Ship captains came to possess vast estates in the several colonies. Carolina, Georgia, Maryland, Pennsylvania, New Jersey, and New York all gave headrights in various forms and sizes.

PLYMOUTH AND THE MASSACHUSETTS BAY COLONY

In 1620, the Plymouth Company financed the settlement of the Pilgrims on the territory of the Virginia Company; but the colonizers missed Virginia and landed near Cape Cod instead. The Crown granted the Plymouth Company a monopoly in this area in 1621. Under the leadership of William Bradford, the Pilgrims saved their profits from fur and lumber, and in 1641, they bought the Plymouth Company, the corporation that owned the land.

In 1628, a group of Puritans with governmental permission had settled at Salem in Massachusetts Bay Colony. Other Puritans secured a monopolistic charter for the Company of Massachusetts Bay in New England in 1629. Neither the Plymouth nor the Massachusetts Bay colonies earned any appreciable profits for their shareholders, and so the original capitalists readily sold out to the colonists. The Puritans gained control of their company and operated it for the glory of God rather than for profit. This devotion may have been easier to achieve since the company lost money. In each of the northern colonies, the immigrants used land as the basis for a biblical commonwealth.

At first Plymouth settlers had been indentured servants of the company on company land. After the Pilgrims purchased the company charter, they soon operated their land system on the pattern developed in Massachusetts Bay. From the first, the Puritans of Massachusetts Bay owned their company and its land, and they parcelled out land in small pieces to members of the church in good standing.

The Puritans retained their own distinctive version of ownership and operation until about 1762.

COLONY OF GEORGIA

Only one other company received a charter to colonize in the New World. Under the guidance of James Oglethorpe, the Crown granted a charter to a company to settle Georgia in 1732. The charter, however, provided that the colony and its land would revert to the Crown in 1753. The Crown used its Virginia experience in the charter. Oglethorpe formed the company mainly as a refuge for debtors who were given the chance to go to Georgia instead of an English prison. The company gave 50 acres to every colonist, thus copying the Virginia headright. Colonists who paid their own passage and who brought a family of six received 500 acres of land. In 1752, the company gladly surrendered its colony to the Crown which continued the former land policy.

From beginning to end, company land ownership had failed to work, mainly because the companies could not enforce their ownership. The colonists, in turn, refused to let the system work as long as they had any chance to force the companies to give land to individuals. Except for the Georgia experiment, the company system of land ownership and settlement had ended by 1629.

Two different systems of land disposal and tenure operated in the English colonies by 1629. Southerners embraced the headright system with scattered settlements and an early reliance on cash crops. Northerners developed small holdings, compact village settlements, and a diversified agriculture with emphasis on animal husbandry. The middle colonies followed in general the concepts and pattern of Virginia.

OTHER METHODS OF OBTAINING LAND

By the 18th century, most farmers in the southern colonies obtained their land through purchase from proprietors, the Crown, or other landowners. Virginians also acquired most of their land in this way after 1700. Maryland ended the headright in 1663 and thereafter only sold the land. By the 18th century, the Carolinas and Georgia mostly sold land as did the proprietors of the Middle Atlantic colonies. Purchasers usually had to buy in parcels of no less than 50 acres.

From time to time the Crown gave large pieces of land to various

Farm scene, *ca.* 1777. Sketch, New York, 1848. (From B. J. Lossing, *A Pictorial Field Book of the Revolution,* Vol. 1, New York, 1852)

individuals. Those proprietors then had to devise ways of either using or disposing of their land. Efforts to establish a sort of feudal tenure failed to work satisfactorily. The proprietors found it more profitable to sell or dispose of their land rather than try to maintain American colonists in tenancy. Proprietors received land from the Crown in Maryland in 1632, New Jersey in 1664, New York in 1664, Carolina in 1666, New Hampshire in 1679, and Pennsylvania in 1681.

In Maryland, George Calvert, Lord Baltimore, intended to create a refuge for Roman Catholics. He granted a headright of 100 acres for each settler and an additional 100 acres for anyone the settler brought. The several proprietors of New Jersey gave headrights of 150 acres of land to the settler and 150 acres for anyone he brought to America. In the Carolinas the proprietors used a complicated headright system. New Hampshire followed the New England system of settlement and largely ignored the proprietor, John Mason. William Penn, then the second largest landowner in the world, tried to sell most of his 47 million acres. He asked £2 to £5 sterling for every hundred acres. But Penn also gave away large tracts. He granted headrights of 200 acres to Pennsylvania settlers who paid their own way, with headrights of 50 acres for every additional immigrant.

In New York the English continued the policies of the Dutch. The Duke of York followed the policy of lavish grants begun by the Dutch. Prior to the 18th century, a headright grant in New York usually came to no less than a thousand acres. The duke, later king, also sold land. During the 17th century and sometimes into the 18th century, the proprietors in Pennsylvania, New York, and all the southern colonies tried to collect quitrents of cash to take the place of feudal work and commodity obligations. They may have been regularly paid in the more heavily settled areas, but farmers soon became discontented with these feudal obligations.

THE DUTCH IN NORTH AMERICA

Following the explorations of Henry Hudson in 1609, Dutch traders went to the Hudson River and traded for furs with the Indians. The Dutch established a post on Manhattan Island by about 1613. The Netherlands' Estates-General chartered the Netherlands West Indies Company in 1621, giving it authority to colonize nearly everywhere in America and Africa. The company set up the first permanent settlement on Manhattan in 1623. As in the English colonies to the north, settlers came as employees of the West Indies Company which kept title to the land. Few migrated from Holland on those terms.

In 1629, the Netherlands West Indies Company began issuing Charters of Privileges to Patroons. These charters gave individuals title to 8 linear miles on either side of the Hudson, or 16 miles on one side for an indefinite distance into the interior. To receive the grant, the patroon had to bring over 50 families within a period of four years. In contrast to the policy in the English colonies, the patroons had to buy the land from the Indians. The patroons became landlords, and their settlers semifeudal tenants. The system created a few large, sparsely populated estates.

In 1640, the Netherlands company allowed the creation of smaller patroon holdings of 200 acres for those who could bring over five men, and in 1650, individual settlers could get land and stock for a fixed rent with the understanding that they pay for the stock at the end of six years. The Indians put continuing pressure on the Dutch. Dutch government in America lacked both efficiency and humanity, so the English conquest of the colony in 1664 came almost as a relief to many of the Dutch farmers. The English continued the policy of granting large estates, however, and by 1764 about 30 families held most of the province outside the Iroquois country. Three estates contained over a million acres each; other estates had less but they had the best land. Farmers had largely fallen to the status of tenants by the eve of the Revolution.

TOWN SETTLEMENT SYSTEM

Individual farmers sometimes received land grants in the New England biblical colonies, but these governments made grants infrequently for comparatively small parcels and always for services rendered or to be rendered. Farmers occupied most of New England in farm villages directly or indirectly under the guidance of the church. Beginning in 1628, Massachusetts settlers had to secure town grants from the General Court which served as both legislature and appellate court.

In Connecticut and Rhode Island, the original proprietors secured the land from the Indians and then had the grants confirmed by the Crown. The governments in both colonies tried to distribute the land to farmers with acceptable religious or political views. In Connecticut and slightly later in New Haven, the church leaders exercised considerable control over settlement and land distribution in contrast to civil authorities. The theocratic system led to confusion of titles and boundaries. In 1639, New Haven ordered settlers to secure government approval first as done in Massachusetts. Connecticut and Rhode Island legislatures passed similar laws.

The Crown granted Maine and New Hampshire as proprietary colonies to Ferdinando Gorges and John Mason in 1622 and 1629, respectively. They gave large estates to individuals within their territories, but farmers already there soon secured effective control of land disposal and titles. The farmers took over John Mason's holdings when he died in 1635. Settlers from Massachusetts moved into New Hampshire under the general pattern used in Massachusetts. By 1641, these farmers had become numerous enough that the rule of Massachusetts Bay was recognized in New Hampshire. Similarly, the Maine settlements came under control of Massachusetts in the 1650s, and the town settlement system prevailed. Maine formally became a part of Massachusetts in 1668, and Maine settlers sent delegates to the Massachusetts General Court in 1669. Gorges still remained an important landowner in Maine, but Massachusetts bought up the claims of his heirs in 1677.

In 1679, the heirs of John Mason sold their claims in New Hampshire to the Crown which separated New Hampshire from Massachusetts in 1680. Nevertheless, New Hampshire settlers tended to follow the Massachusetts town system of settlement until the mid-18th century. By a variety of routes, all of New England embraced the Massachusetts town system of settlement in the 17th century and continued it successfully until the French and Indian War. The land policy in New England attracted many farmers and the colonies grew in spite of the unfavorable climate and generally poor soil.

TOWN GRANT METHOD

In the New England system, the town grant method transferred land from the state to individuals. The New Englanders aimed to foster group settlement and propagation of the faith without securing any particular profit for the state. When 20 or more men had decided to move, they sent a petition to the General Court asking for a grant. If the General Court agreed, it appointed a committee made up of the petitioners and members of the General Court.

This group, called the Viewers, would then select a site for the new settlement. The land chosen had to adjoin the land of some other town in order to preserve compactness of settlement. Having made their selection, the Viewers then sought the approval of the General Court. If it approved, the petitioners then became the town proprietors, and title to the land passed from the state to the town with definite rules for its disposal.

The town proprietors owned the land but had to distribute it. Newcomers to the town had to be included in the group of proprietors until a total of from 60 to 80 family heads had been added. The exact number depended mostly on the size of the land given to the town. The plan of distribution varied little during the century. Each proprietor received one town lot and several out lots. The town lots were laid out in sizes of equal value but not necessarily of equal extent. One town lot was kept for the church, one for the minister, and one for the school. The out lots differed in size and in the number given to each farmer.

Several considerations determined the division of the farmland. The amount of taxable property a settler brought into the town often influenced the amount of land he received. Those with the most property received the most land. However, a poor man with a large family usually received a large piece of land. In a few towns the size of the family chiefly determined the amount of land distributed to individuals. In either case, the disposal method tended to reward the Puritan virtues and attract desirable settlers. Presumably the man with the most property worked the hardest and the town needed hard workers. On the other hand, the prosperity and protection of the town required a large and growing population.

The corporate entity, the town, also kept some land. These reserves fell into three categories: the village green, the common field, and other undivided land. The village green, located in the center of the village, was to endure forever as a place of repose and as a drill field for the militia. The common served as a pasture for the farmers who could run specified numbers and kinds of animals. The proprietors only had rights to use of the land. The size of the farmer's landholdings largely determined his rights in the common. Sometimes the community cultivated the land, and as time went on the town generally distributed the common to individuals. The other undivided land was held in reserve and distributed as the town population grew either through natural increase or through immigration. In the latter 18th century, this undivided land often fell into the hands of land speculators.

The size of a New England farm varied greatly, but holdings apparently ranged from 12 to 20 acres in the original distribution, not counting any rights in the common. The farming fields were distributed in strips or plots, perhaps three to five, so that each farmer received a share of each quality of land. The farmers lived in the centrally located village with their fields round about. Consolidation of the separated fields began in the latter half of the 18th century, but it was not completed by the end of the Revolution.

The New England system of land disposal compared favorably with the systems used in the colonies of the South and the Middle Atlantic. The New Englanders gave the land away generously. In contrast, the headright system practically ended in the other colonies by the late 17th century. Compared to most of the other colonies which followed the Virginia system, the New Englanders secured compact settlements with the attendant advantages of church, school, and defense all in one package. The town system also sharply reduced the opportunities for land speculation and prevented the development of huge estates or plantations until much later in history.

SHORTCOMINGS OF THE TOWN GRANT SYSTEM

The town grant method did have some shortcomings. Eventually the town no longer welcomed newcomers and no land remained for distribution. Nothing prevented any individual from being a petitioner time after time for a proprietor did not actually have to reside in the new town. Thus most towns had several absentee proprietors. These men did not accumulate vast estates in single large blocks as in New York or Virginia. The New England landholder had large estates spread among many towns in small parcels. Antagonism eventually developed between the landless laborers and the absentee landowners.

By 1700 at the latest, land engrossment in New England took the form of speculation in farmland, which increased as religious fervor declined and clerical sanctions came to have less impact. Ultimately the town grant system collapsed, partly because of speculation, but largely because changing methods of farming better utilized larger farms in consolidated holdings. Massachusetts abandoned the system of town grants in 1762 when the Commonwealth sold its last lands to individual investors. In part, Massachusetts needed the immediate revenue to finance the French and Indian War.

In the 17th century, many towns had already begun selling some of their undivided land to individuals, and many of the common lands had also been sold. The revenue went chiefly to defray war debts and to provide for schools and libraries. Beginning in 1725, Massachusetts and Connecticut began selling some township grants to individuals for resale. Many of the speculators were prominent officials. They defended the policy change as an effort to speed settlement by turning promotion over to speculators. Soon speculative booms occurred in New England: one boom between 1730 and 1740, and another later in the century.

New England agriculture expanded comparatively rapidly during the 18th century.

MILITARY BOUNTY METHOD OF LAND DISPOSAL

The military bounty was another method of land disposal which grew in importance throughout the 18th century and became even more important in the 19th. Military land bounties apparently began in 1646 when Virginia gave 100 acres of land to the commander at Middle Plantation. Following King Philip's War in 1677, Massachusetts devised the military town with special grants and privileges to the settlers. In 1679, Virginia gave Lawrence Smith and William Byrd land grants provided they settled exposed areas with armed men. Virginia extended this privilege to everyone in Virginia in 1701. Many colonial legislatures began laying out townships around 1713 and giving them to veterans of the earlier wars. In 1733, Connecticut gave nine townships to officers and soldiers, or their heirs, who had served in King Philip's War. Most veterans sold their shares to speculators. In 1755, Pennsylvania offered grants to those who would join an attack against the French. The Proclamation of 1763 gave soldiers of the French and Indian War from 50 to 5,000 acres according to rank. The Crown issued warrants which the possessor could use for any Crown land. Although farmers and speculators apparently made little use of these military grants, these acts set precedents for American grants during the Revolution and later grants by state governments involving millions of acres.

The results of colonial land policy varied. First of all, land policy had little influence on the crops the farmer chose to grow. Land policy did have a profound influence on the number of settlers

Farmhouse, *ca.* 1780, Cherry Valley, New York. (From B. J. Lossing, *A Pictorial Field Book of the Revolution,* Vol. 1, New York, 1852)

who migrated, their territorial distribution, and the nature of the evolving labor system.

For example, conditions of land tenure in New York through the colonial period encouraged immigration into Pennsylvania or further south. New England, with its restrictive systems, was too forbidding to attract much migration from other regions. In New York, potentially one of the best farming areas, settlement proceeded slowly. In part, farmer reluctance to migrate resulted from problems of securing title to land. The New York farmer had to buy the land from the Indians, then secure the approval of the governor, finance a survey, and finally secure a title from the governor and council. From Indians to governor, the farmer encountered fraud and bribery at every step. This favored the larger operators who could afford the chicanery. Squatters eventually became the tenants on some large landholdings. Large groups, such as the German immigrants of 1710, could get grants, but large group movements were not the typical way of settling. Consequently, pioneers turned to other places with easier, more generous land policies.

The flood into Pennsylvania began around 1710. Most of the settlers came as indentured servants; and after serving their time, they went to a frontier area. Recently freed servants could seldom afford to buy land from speculators or even from the colony; so for the most part, they squatted on the land and hoped to make enough money to secure title before being evicted. In 1726, over 100,000 settlers in Pennsylvania lived on land to which they had no title. Faced with a squatter population of this magnitude, the speculators and the colony recognized the right of preemption with later payment around 1750. Pennsylvania charged £10 sterling for a hundred acres until 1738 when the colony raised the price to £15 a hundred. In contrast, Maryland charged only £5, and in the Shenandoah valley of Virginia the speculators charged even less. In a short time, farmers began to move in large numbers out of Pennsylvania into Maryland and Virginia.

Meanwhile the tidewater areas of Virginia and the Carolinas began to fill up in the early 18th century, and the Piedmont attracted more and more settlers. Farmers slowly began to move into and beyond the Appalachian Mountains. This hesitant migration of farmers from Pennsylvania to the Carolinas virtually ended with the French and Indian War (1754 to 1763). Most Indians supported the French, and pioneering became fairly dangerous. The victory of the English by 1761—ratified by the Treaty of Paris in 1763—left the Indians weakened and without allies.

At the close of the French and Indian War, Britain came to be the only power from the Atlantic to the Mississippi. New land speculating companies appeared, and a new surge of farmers began westward. This movement of farmers and some other alarms caused an Indian uprising, Pontiac's Rebellion, late in 1763. The Indians devastated frontier farming settlements. The British government, with some colonial assistance, put the Indian rebellion down. The British had just emerged from an expensive and frightening war and sought no more confrontations with the Indians.

CHANGING BRITISH COLONIAL POLICIES

The British concluded that land seizure, treaty frauds, and migration of farmers had brought on the Indian war of 1763. The British government, which then had no unified land policy, decided to devise one. Meanwhile land theft and land deals with the Indians were stopped. As early as 1761, the British government had ordered an end to further land purchases in the West by the colonial governments or their residents. No one paid any particular attention to the order.

On October 7, 1763, King George III issued the Royal Proclamation of 1763. Four new American colonies with formal boundaries and governments were created: the British West Indies, the Province of Quebec, the Province of West Florida, and the Province of East Florida. The definition and boundaries caused the most concern for farmers, speculators, and traders in Virginia and Georgia because the proclamation deprived these colonies of territory they thought belonged to them.

The king then drew a proclamation line that ran along the crest of the Alleghanies and gave the land west of the line to the Indians. The proclamation declared: "We do strictly forbid, on pain of our displeasure, all our loving subjects from making any purchases or settlements whatever in that region." The proclamation ordered settlers already west of the line to return to the east, ending all immigration into Indian lands. Pioneer farmers no longer had access to all the land of America. The British government had devised a land policy which they had to uphold and which they bequeathed to the Americans.

Not until 1887 did Americans abandon the idea of reserving certain land for the Indians. The Indian reservation concept (which appeared again in 1934) continued as policy, although the size of the reservations diminished markedly. The British soon acknowledged the proclamation line and the land reserved for the Indians to be only temporary. New lines and, in effect, new reservations were to be negotiated with the Indians.

The Crown intended to halt American speculation and migration until it worked out an imperial land policy that emerged in 1764 with later modifications. Few settlers moved from west to east simply at the wish of the Crown. To force farmers off their farms something more than voluntary enforcement had to be employed. Farmer attitudes proved hard to change. Under the old ways of getting a farm lay certain economic realities which also failed to change. American agrarian antipathy to Britain increased with every effort to enforce the new policy.

The base of an independent agriculture had developed in America by 1763. Farmers had slowly forced on each colonial government a policy which facilitated individual acquisition of land. First companies, then proprietors, relinquished land to those who farmed it. The pattern of distribution to private owners varied from colony to colony, without much plan and certainly no uniformity. Suddenly in 1763, the Crown and Parliament sought to impose an imperial land system on the New World.

CHAPTER TWO

LAND AND THE REVOLUTION ⚔ 1763–1787

AFTER 1763, the American colonies continued their several methods of land disposal within the framework of the Proclamation of 1763. But they could not control squatters. In Pennsylvania and Virginia, governmental leaders began working for the repeal of the proclamation. The English Parliament enacted supplementary laws in 1764 to enforce and alter the proclamation.

For enforcement, the Act of 1764 established a Northern and a Southern Indian Superintendency. William Johnson, married to a Mohawk, became Superintendent for the North; and John Stuart, a blood brother of the Cherokee, became Superintendent for the South. Although the superintendents received very little money, they had to regulate both the Indian trade and land settlement with little help. They were to enforce the Proclamation of 1763 by making sure that no individual purchased land west of the line and by moving settlers west of the line to the east of it. Stuart moved farmers, who took the first opportunity to move back. Both Johnson and Stuart kept sales by the Indians to a minimum, but settlers in the South increasingly subverted the law by engaging in long-term leases with the Indians.

The law also charged the superintendents with keeping undesirables out of the Indian trade and with regulating that trade to keep the Indians at peace. Trade was to be carried on by license only. No rum could enter the trade, and certain types of rifles were not to be sold to the Indians. Every trading post was to have a blacksmith to repair items the Indians could not fix themselves. The superintendents licensed traders, but Johnson and Stuart lacked the resources to carry out the other provisions of the law.

In addition to everything else, the superintendents had to negotiate new boundaries with the Indians to alter the Proclamation Line of 1763. The superintendents did this, and in the process not only effectively broke the Proclamation Line but virtually assured western migration. If the line could be broken once and renegotiated, it could be broken again and again. In any case, the new treaty lines were difficult to locate, did not join up properly, and generally pushed Indian boundaries far to the west. Farmers moving west could usually find enough land within the newly opened areas, and they could claim that they could not find the lines anyhow. Neither Johnson nor Stuart could effectively patrol the lines. The true farmer-settler had some security of tenure because large grants were prevented under the act. The squatter could make his deal with the Indians without much fear that a land speculator or the colony would eventually deprive him of title.

RENEGOTIATION OF THE PROCLAMATION LINE

As early as 1763, Stuart negotiated a treaty with the Cherokee at Augusta which set the line in the Carolinas and Georgia. In the Treaty of Hard Labor in 1768, he had the line extended to the confluence of the Great Kanawha and the Ohio rivers. In 1768, Johnson extended the line north from the Ohio River through New York and Pennsylvania at the Treaty of Fort Stanwix with the Iroquois. There the Iroquois surrendered their shadowy claims south of the Ohio.

The land thus surrendered by the Indians could be purchased by land speculators. As settlers moved into the newly opened land and beyond, land speculators demanded that the line be moved even further west. By the Treaty of Locahber in October 1770, the Indians agreed to have the line moved in Virginia. This renegotiation stimulated new activity by the land companies, old and new. Land companies seeking grants essentially took up where the older proprietors and companies of the 17th century had left off. The limited company grant provided the Crown an orderly way for disposing of land and securing some revenue from it. In 1732, the last such grant created the colony of Georgia. The establishment of Georgia encouraged the view that the acts of 1763 and 1764 might be relaxed gradually and new grants made. Company grants offered the settler-farmer some guarantee of a regular and legal title to their farms.

THE GRAND OHIO COMPANY AND VANDALIA

In December 1768, the process of breaking the line of 1763 was well under way; so Samuel Wharton of the Indiana Company began to consolidate his group with other companies. Absorbing rivals seemed the only sure way of getting a large and profitable grant. Apparently the Crown preferred only to deal with large grants. The Indiana Company merged with the old Ohio Company and individuals such as George Washington and some British statesmen. The Colonial Secretary suggested that the company apply for a grant as a colony. This was done under the name of the Grand Ohio Company in December 1769.

The new company asked for a colonial grant running from the forks of the Ohio to Greenbriar and west to the mouth of the Scioto. This colony would have contained 20 million acres. (Penn had received 47 million.) The sponsors of the Grand Ohio Company proposed to call their colony "Vandalia" in honor of the Queen of England who believed she descended from the Vandals. The company proposed to pay the Crown a quitrent for 20 years. The projected colony started its legal way through Parliament and the various councils, but it did not clear the bureaucratic hurdles until 1772. A charter to the proprietors, a simple formality, would have established one more colony in America. But the British were already having troubles with the colonists by 1772, so officials held up the charter until conditions quieted in America. The troubles only increased, and the outbreak of war in 1775 ended forever the prospects of Vandalia. The Queen must have been disappointed.

Several other schemes flourished before the Revolution but none came as close to actuality as the Vandalia colony. All the other companies sought Crown grants, and all also illegally bought land from the Indians west of the Proclamation or treaty lines. None of the efforts succeeded, and the deficiencies of British imperial land policy were shown. The Crown might have supervised the movement of farmers; it might even have regulated the sale of land; but it could not absolutely halt all westward migration with anything less than a massive military patrol of the frontier.

TRANSYLVANIA LAND COMPANY

Troubles between the Americans and the British were already near the breaking point when the last important colonial land scheme developed. Its failure provided interesting precedents for the emerging American national land policy. In 1774, Richard Henderson and several associates formed the Transylvania Land Company which tried to secure title to 2 million acres of land west of the mountains and between the Ohio and Cumberland rivers. The land, well beyond the treaty line, belonged primarily to the Cherokee. Henderson apparently ignored the British altogether and devoted his attention instead to the Indians, the colonies, and later to the Continental Congress. Henderson got around British law by leasing the land from the Cherokee, something farmers had been doing for some time. Various settlements began in the proposed colony of Transylvania, and Henderson made leases with the Indians at Sycamore Shoals in March 1775. This arrangement amounted to a purchase. Henderson hoped to get his colony established so quickly that it could not easily be broken up. His fears of disruption centered more on Virginia and North Carolina than on Great Britain.

War broke out April 19, 1775, and Henderson turned his attention to the assembling Second Continental Congress. He sought to have the Congress recognize his Transylvania as a fourteenth province, but representatives of Virginia blocked his efforts. The Congress decided not to create personal and private domains and thus ended one aspect of British colonial policy. To meet the needs of farmers in the western area, Virginia created the County of Kentucky on December 7, 1776, and the Transylvania scheme ended.

EMERGENCE OF AMERICAN LAND POLICY
DURING THE REVOLUTION

The existing states and the Congress had become heirs of the Crown and Parliament. The Americans needed to formulate a land and government policy for their new domain. Details of the land policy were quickly settled to meet the exigencies of war and the national need for troops and revenue.

The War of the Revolution was no trivial enterprise. American farmers and merchants had dared to oppose the most powerful nation in Europe. The contest went on vigorously for seven years. The Americans began the war with no national army, no treasury, and very little bureaucratic machinery. But the Americans owned America, no small resource. Americans had induced enlistments since 1646 by offering land to soldiers. The assembled Congress, possessed of no treasury and with no powers of taxation, found land in lieu of money a particularly attractive solution to the problems of carrying on the war. Oddly enough, the Congress offered the first bounties of land to the enemy rather than to loyal Americans. In August 1776, Congress offered land to deserters from the Hessian troops employed by the British. The bounties alone cannot explain the high rate of desertion by the Hessians. But, at the end of the war, Germans who had deserted or refused to go home after capture may have numbered some 4,000 or more out of a force of nearly 20,000. Free land and a farm proved very attractive to these German mercenaries.

LAND BOUNTIES

The United States had to have a national army, the Continental Line. On September 16, 1776, the Continental Congress offered land bounties to troops enlisted for the duration of the war. To induce enlistments, it provided for the warrants to be issued only at the end of the war. Delayed payment presumably would guarantee the length of enlistment. Grants under the act varied according to rank: 500 acres to a colonel, less in regular downward steps to other officers, and 100 acres to enlisted men. In 1780, the Congress increased the size of the grants, although it did not have any land with which to redeem the warrants at that time.

The states also offered bounties to induce enlistments in the state armed forces. The Virginia legislature passed the earliest such act in 1779. Virginia offered 100 acres to each soldier at the end of the war, and the grants to the officers equaled those offered by the Congress. Virginia had claim to territory, and soldiers could use the warrants to secure land in Kentucky and Ohio. Other states offered bounties, even though several of them had little or no land to honor their obligations. In 1780 and 1782, Virginia increased the size of her bounties. Massachusetts, New York, Pennsylvania, Maryland, North Carolina, and South Carolina also offered military bounties. In 1780, Virginia set up the Military Reserve in Ohio between the Scioto and Miami rivers to fulfill state needs. In 1781, Virginia offered to honor the warrants issued by landless Maryland, and Virginia kept her promise after the war.

STATES CEDE LAND TO CONGRESS

Congress needed land to meet its bounty obligations as well as sources of revenue independent of the uncertain state contributions to the national treasury. The congressional dilemma was eased by the land speculation companies whose private objectives happened to coincide with national needs. Prominent politicians and soldiers such as Robert Morris, Benjamin Franklin, and George Washington were members of land speculation companies so the land speculators exercised considerable influence in the Congress and in the state legislatures. Both the Vandalia Company and the Illinois-Wabash Company still existed when the war began. Both had made illegal purchases from the Indians, and their officers knew that the individual states would not recognize their claims, Virginia, which had claims in the interior and its own speculators' interests to consider, would not validate the holdings of outsiders. The company officials wanted the states to cede their land to the Congress in the hope that it would ratify the claims of the speculators in return for some monetary payment. Congress also wanted land to sell.

In 1780, New York gave up her tenuous claims to western territory; the same year Virginia offered her lands in Ohio and the Northwest to Congress. Virginia provided only that Congress not recognize private purchases made in the territory prior to the cession. Congress, eager for land, accepted Virginia's proviso; and Virginia ceded her land to Congress in 1781, keeping only the Military Reserve to pay her veterans. Massachusetts followed in 1784 with a gift of her unclear claims, and Connecticut surrendered her claim to a strip extending to the Mississippi (except for a small portion in Ohio) in 1786. Connecticut kept title to this portion called the Western Reserve, while surrendering any political jurisdiction. All states surrendered their sovereignty in the areas involved, including reserves.

The states south of Virginia surrendered their land a little more slowly. Probably they delayed more because they had more land to give up. In 1783, North Carolina withheld a small military reserve on the Cumberland River and turned the rest of the claim over to land speculators. Having taken care of her own by giving them title to what was to become Tennessee, North Carolina then turned over its sovereignty to Congress in June 1784. But the same year North Carolina took back control because speculation dealings had caused an Indian war. John Sevier and other land speculators attempted to have Congress recognize their domain as the state of Franklin. North Carolina settled the Indian war, crushed the projected state of Franklin, and then ceded Tennessee to the United States again in December 1789. Congress accepted the gift in February 1790.

In 1787, South Carolina had ceded a small strip of land along its northern boundary to Congress. Of the states having western land, only Georgia still held its land in 1789, when it sold some land to private speculators. In 1795, the Georgia legislature turned over most of its western land to its own members under the title of the Yazoo Land Company. The Supreme Court in one of its earliest decisions sustained this fraudulent deal on the grounds of sanctity of contracts. In 1802, Georgia relinquished political control over her western country which embraced most of the future states of Alabama and Mississippi.

Bit by bit the United States secured an imperial domain and a sizable amount of land to sell or give away. For the most part the Congress hoped to sell the land. The end of the war in 1783 made revenue a less pressing consideration. In 1788, the adoption of the Constitution ended the pressing need for an independent source of income. Nevertheless, both imperial and land policies had been agreed on well before the Constitution was adopted.

THE LAND ORDINANCE OF 1785

Acting for the Congress of the Confederation, a committee met through 1785 to work out a profitable program of land disposal for the almost indigent Congress. The Congress sought an orderly land survey to account for all parcels and to collect all fees and a system of distribution which would yield maximum returns to the treasury. The resulting Land Ordinance of 1785 cunningly created all the desired features except a horde of settlers with ready cash.

The Land Ordinance of 1785 provided that the land should be surveyed in square blocks, which were called rectangular, even though the curvature of the earth actually formed trapezoids. The survey started from base lines and meridians, and a series of these eventually came to be designated across the country. Measuring from the base and the meridian, surveyors laid out range and town lines at six-mile intervals. The ranges, or blocks, ran from east to west, and the town squares north to south. Each township ran 6 miles on each side, and within the township surveyors marked out 36 mile-square sections.

Each section contained 640 acres which could be surveyed into quarters of 160 acres each. Congress reserved Section 16, near the middle of each township, for the support of schools. Section 16 was to be sold by the state or territory to produce money for education. The orderly laying off of the land came before its sale although not necessarily before settlement. The land was described by numbers that indicated the town, the range, the section, and the subdivision within the section. This part of the act was copied for later territories of the United States.

PROBLEMS OF THE LAND ORDINANCE

Historians have often praised the wisdom and generosity of the Committee on Public Lands which drew up the Land Ordinance of 1785. Actually the act carried several illiberal features, some of which Congress later corrected. For example, the purchaser of the land had to buy it at a Seat of Congress instead of where the land was located. Land was sold at public auction with a minimum bid of a dollar an acre in hard money. The one dollar minimum bid only started the bidding. Most land sold at auction for far more per acre. High bids pushed the price far beyond the means of the farmers on the land. Every other township had to be sold in one parcel, which came to a minimum bid of $23,040 for alternate townships. The purchaser could bid for nothing less than one section of 640 acres in the other townships.

These provisions favored wealthy speculators over farmers who could seldom make the trip East. They could not bid for land they already farmed against a speculator with ready cash. When a farmer might accumulate $50 in hard money in a lifetime, a minimum bid of $640—the least amount allowed by law—exceeded the resources of most farmers. Furthermore, 640 acres was at least eight times more land than any one man could farm.

The farmers on the land protected themselves by uniting to prevent the sale of land they had already settled. Purchasers other than the pioneers could be intimidated unless the sale took place at a Seat of Congress such as Philadelphia or New York. Sale of land was not a land-office business, although pioneer farming continued to expand. Well-armed, truculent squatters did not increase respect for law or facilitate the orderly disposition of land. Furthermore, the sort of settlers attracted to western lands under these conditions seldom turned out to be the most upstanding citizens of their community.

Rough-and-ready frontiersmen, however admirable, could not be expected to pay taxes or live orderly lives in the view of eastern businessmen and politicians. The evidence seemed to support their estimation. Sturdy, law-abiding, citizens could not be expected to go west, buy land, and pay their taxes unless they received some inducement to move and the promise of a regular government. And so it fell to the Congress of the Confederation to organize that part of the American empire which the states had ceded to it.

THE OHIO COMPANY

The governmental system and imperial policy actually grew out of the needs of a land speculation company chartered in Massachusetts. This Ohio Company was founded in 1786, and after some maneuvering it persuaded the Congress to sell it 5 million acres of land in Ohio. Of this total, the Ohio Company surrendered some 3.5 million acres to the Scioto Company which had many prominent members of Congress on its list of subscribers. The Scioto Company modestly chose not to ask for land for itself directly. Instead the members of Congress accepted the special terms for the sale of land to the Ohio Company which had fewer congressmen in it. Appearances did matter, after all.

The Ohio Company proposed to buy only 1.5 million acres for

itself. It proffered Massachusetts Certificates, Continental Currency, and other paper money instead of the hard money required by the law of 1785. The amount offered came to about 8 cents an acre in hard cash. The Congress accepted this proposal. The promoters of the Ohio Company wanted the Congress to assure law and order in the area. Such assurance would help the company attract settlers from New England, widely assumed to be a more law-abiding area than most. The company wanted God-fearing, debt-paying settlers.

THE NORTHWEST ORDINANCE OF 1787

The Congress obliged by passing the Northwest Ordinance of 1787 which represented a detailed consensus on U.S. imperial policy consistent with the yet unfinished Constitution. Operators in the Ohio Company, such as Nathan Dane, Rufus Putnam, and Manasseh Cutler, wrote the ordinance and assisted in pushing it through the Congress. Although evolved in this curious way, the act established a far-sighted and liberal imperial system for the United States. For the first time in history, an imperial power set up—in advance of any serious trouble—a method whereby the colonial possessions would assume full and equal status within the mother country.

The act specified how to increase the empire and provided for government from the beginning of settlement. The Old Northwest was to be made into not less than three nor more than five territories, and these territories were eventually to enter the Union as states. The law tentatively set boundaries of the territories and thus of the projected states. The Northwest Ordinance provided an executive government during the territorial period consisting of a governor and three judges appointed by Congress. When the population of any of the territories reached 5,000, the people of the territory could have a legislature. When a territory (one of five major divisions) had 60,000 people, they could write a constitution and apply for admission to the Union on equal terms with the other states. Each state constitution had to include a bill of rights and had to prohibit slavery in those states formed from the Northwest Territory. The new states were prohibited from passing laws of entail and primogeniture.

The provisions against entail and primogeniture bore directly on land policy and the lives of future farmers. Laws of entail prevented a landed estate from being broken up either by sale or by inheritance. Under entail such a landholding had to revert intact to the state should the last owner have no heir. Laws of primogeniture required that the oldest surviving son or direct male heir receive the entire estate intact. These laws fostered the development of an agricultural aristocracy. The laws had developed in feudal Europe to achieve that aim and had restricted the development of farming by preventing the free exchange of land. The older states had abolished these leftovers from feudal England and proposed to prohibit the adoption of such laws in the new states. Thus the Ordinance of 1787 helped to assure a flexible land system for American farmers.

The political arrangements set forth in the Northwest Ordinance of 1787 also encouraged farmers to take up new lands. They did not have to fear tyranny nor have to take up arms against the mother country to protect their property. As the agrarian empire expanded, farmers increased their wealth and the wealth of the nation.

Together, the Land Ordinance of 1785 and the Northwest Ordinance of 1787 set the pattern, and the policies laid down by the Congress of the Confederation easily fitted into the new system. Both ordinances reflected what the majority of Americans wanted: land, law, and order.

CHAPTER THREE

FARMING: METHODS AND TOOLS ❦ 1607–1783

AMERICAN FARMERS inherited two quite different agricultural traditions: those of Western Europe which contained a vast accumulation of empirical lore, and those of the American Indians and the African slaves. The Europeans brought a post–Iron Age technology along with the plants and animals developed during the Old World Mesolithic and Neolithic ages.

The Indians had worked out methods to suit the American climate, soils, and native crops. As skilled plant breeders, they contributed the high-yielding corn and potatoes as well as beans, varieties of squash, and tobacco. The Indian technology corresponded roughly to that of the late Neolithic period of Europe. Indian tools of bone and stone differed little in use or design from the metal implements of the Europeans.

The Iroquois and Muskhogean practiced a sophisticated and successful garden agriculture without plows. The Indians of the Americas had no draft animals. The first plows of the Old World may have been pulled by humans before they used draft animals. New World crops did not require as much soil breaking as wheat and rice.

The African slaves also introduced plants and methods developed in their own advanced garden agriculture. The several sorghums were the main plants. Later on men systematically brought sorghums to America, but apparently the African slaves made the first introduction. They also introduced various spice plants and may have brought rice to America.

The dominant Europeans inherited a tradition of wasteful and extensive farming in contrast to the more meticulous garden agriculture of the Indians and Africans. Most Europeans in the 17th century came from cities and towns and had little agricultural experience. These circumstances may explain the readiness of the first settlers to adopt the crops and the methods of the Indians. They could make many of the simple tools and implements they used. Farming by the immigrants started hesitantly.

AGRICULTURAL PRACTICES AND EQUIPMENT

Domesticated animals have always made good use of empty grazing farmland and herding has always been a frontier occupation, Old World or New. Animals trailed to market reduced cost of transportation. However, because trail-driven animals became lean and tough, feeding usually took place close to the market. Food storage facilities had to be provided. Barns, sickles, scythes, and root cellars became necessary: hay had to be grown and meadows provided. Stock raisers encountered problems of branding, disease control, slaughter, and meat storage. In addition, the dairy industry, probably older than the meat industry, had to be carried on close to the cities because of the absence of refrigeration or pasteurization. Both animal husbandry and dairying therefore demanded new technological and managerial skills.

The handmade tools of this period were heavy and awkward to use. Hoes, mattocks, spades, picks, and shovels were used. But the labor required was not heavy in comparison to the returns. People could support themselves with supplementary hunting and fishing. Increased human population forced game further into the wilds, so an efficient agriculture eventually required animal husbandry to provide meat.

Wheat was introduced slowly to America, and for nearly a century corn was the principal food grain. Rye or wheat flour was needed for making bread because other grains provided too little gluten. However, wheat and rye not only yielded less than corn per acre, but the bread grains produced a substantially lower seed-yield ratio. The bread grains also required more and different implements and more work.

However, wheat, not corn, could be sold to Europeans. Any serious commercial agriculture included wheat culture which required plows. The Pilgrims were in America for 12 years without plows, and the Viginians only had around 150 in 1650. The farmer had to plow only when broadcasting small grain or growing corn, tobacco, and cotton on a large scale. But a livelihood with some surplus to sell was possible from the American crops without plowing.

Threshing of wheat, rye, or other Old World crops was considerably more difficult and expensive than handling of the New World crops. Farmers and their families could more easily husk and shell corn by hand than thresh wheat by beating it with flails. The wheat still had to be winnowed with sieves, baskets, or hand-driven fans. The wheat was ground into flour with several siftings to get flour fit for bread. The wheat germ caused it to turn rancid and spoil easily. A bushel of wheat computed at the equivalent of four days pay for a laborer reflected pay scales and the value of wheat in foreign trade. The cost of producing Old World grains was extravagant compared to the harvesting of native New World crops.

Only an interest in commercial farming could have produced an increased interest in using European tools. Iron and steel axes were used to clear the forests, but no evidence suggests that the European settlers did it any better or faster than the Indian farmers who usually followed the worldwide Neolithic practice of girdling. The bark was stripped clear in a ring around a tree trunk which killed the tree. Crops were planted in the sunny patches which appeared with the death of the tree foliage. Steel tools gave the farmer an advantage only when he was preparing the wood to sell.

In short, the absence in early years of many 17th-century implements and work animals did not necessarily mean the expenditure of more human labor for the American farmer but probably less. Even without the use of European iron technology, a good living could be made.

Unquestionably, however, New World farmers could grow the commercially useful Old World crops on a more extensive scale if they used plows, harrows, rollers, and similar animal-drawn equipment. The plows were huge wooden affairs with heavy beams, shares, and moldboards. Such plows required two men to operate: one to hold the plow and one to guide the oxen. Four, six, or even eight oxen were needed to drag the plow through the soil at a depth of about three inches. Few implements had metal tips or strips on the cutting or abrasive surfaces, a fact which slowed the work to an average of an acre a day. The wearing parts needed constant replacement. Plows usually had metal coulters and incidental metal parts such as bolts. About 1700, iron became more plentiful and plowmakers usually made share points of iron and nailed iron strips to the moldboards.

In New England, 17th-century towns paid bounties to those who would buy or build a plow for the use of the comunity. This system of community use worked because of the strip farming of the New England town system and the comparatively small size of the farms. In the South, slaves performed the hoe-and-spade agriculture, not unlike the farming of their ancestral homeland. Some southern farmers used plows, particularly when they raised wheat, rye, and, to a lesser extent, rice.

After the farmer plowed, he had to pulverize his land more before seeding. Clod crushers made of rolling logs and harrows with wooden teeth (later iron) helped in the process. The planter then broadcast the seeds of the Old World grains. He usually planted corn and indigo by walking along the furrow, opening a hole with a hoe or mattock, placing the seeds, and then covering them with his heel. In Europe horse-drawn seed drills came into use in the 18th century, but few farmers used them either there or in America. Cultivators, or horse-drawn hoes, gained acceptance in Europe, but in America weeding was done mostly with hoes or small, light plows.

Perhaps caustic European comments on American agriculture resulted from European unfamiliarity with the Indian-African gardening tradition which the American colonials quickly had adopted. Expensive technology was not economically sound when the products had to move in competition with those of farmers nearer metropolitan markets. Substantial increases in European population and urbanization became noticeable in the 18th century. Simultaneously, technological changes in farming slowly started in America.

IMPROVED EQUIPMENT AND ITS INCREASED USE

The 17th and 18th centuries saw an increased use of the plow and other tillage implements and an increase in livestock population. In *American Husbandry* the implements are graphically described. It was noted that they were at least cheap even if clumsy to use. Farmers made their own implements except for the few iron parts professionally made by blacksmiths who hammered out hoe blades, scythes, axes, and all other tools. A few items were imported by the wealthier farmers. The Revolutionary War, with its stimulation of the iron industry, contributed to a general improvement of farm tools and a reduction in their cost.

Plowing with the improved Bull or Carey plows in the North was still extremely slow. Eighteenth-century plowmakers added more iron and designed moldboards for the different varieties of soil. Two or three horses, or four or six oxen, could plow one or two acres a day, using two men plus sometimes a boy. The boy prodded the oxen, and the second man added weight to the plow beam to keep the plow in the ground. Sometimes the plow had a platform on the beam on which large rocks, common in New England, could be piled to hold the plow down. This method had the disadvantage of making the plow difficult to lift and move around obstructions.

During the Revolution, more farmers began using cast iron for parts of the plow, which improved efficiency. In 1797, Charles Newbold patented the first all-iron plow which was cast in one piece. If any part broke it all became useless, but nevertheless his invention represented an advance in construction and design. American farmers have been accused of resisting the use of the iron plow. Some farmers supposedly believed iron poisoned the soil, and this explanation may have been a handy excuse for not using admittedly imperfect tools. However, 18th-century farmers commonly used coulters and shares

Strong plow, *ca.* 1735

Strong plow, *ca.* 1740–1783

German wheeled, 1769

of wrought iron and usually had moldboards covered with strips of metal. Farmers probably rejected iron plows because they saw no economic advantage in their use. However, farmers made the shift rapidly enough when they had adequate economic reasons.

The newer technology had consequences other than allowing one man to work more land. The plow and similar tillage implements hastened soil depletion and erosion as the implements became more effective. The criticized deep plowing of many farmers, resulting from an unusual plow construction, accidentally helped retain the soil. A later generation specifically used deep plowing to retard erosion. Incomplete harrowing, resulting from unusual harrow design and materials, at least did not completely pulverize the soil and hasten its washing.

PIONEER FARMING METHODS

In most places the pioneer cleared about three acres. He put half in corn and half in vegetables. More cleared land would be put mostly into corn which was fed to animals which could be driven to market. Sometimes the new farmer would clear the land until winter and then return to his home. The next year he would return and plant his crops and clear more land for his first year of permanent residence. An acre or more of wheat might be grown to produce a modest marketable surplus. Occasionally during the Revolution farmers put in flax plots to meet the army demand for linen.

Garden agriculture allowed the farmer to get continuing high yields with the use of simple, soil-conserving implements. Virgin soil yielded from 50 to 100 bushels or more of corn per acre. The Indians had sometimes achieved yields well above 100 bushels to an acre! Yields of 30 bushels of wheat were common. New farming methods allowed farmers to crop more land with a greater overall return. With less careful use of the land, yields per acre fell and soil erosion began in earnest.

The small plots and high yields of hoe agriculture encouraged farmers to alternate fields and crops regularly. The shovel plow of shallow cut was usually the first important horse-drawn implement used in newly opened areas. It served as a cultivator to keep down weeds in corn, tobacco, and other crops such as potatoes. The white potato had become a widespread and important crop by the eve of the Revolution. It was seldom grown for market. The ax, hoe, or spade opened the ground enough for potato planting and harvesting.

HARVESTING PRACTICES

The farmer's year ended with the harvest. Southerners commonly cut the tobacco plants, let them lie awhile in the fields, and then stored them in barns or tobacco houses. Or they hung the tobacco outside to dry in the sun or sometimes cured it over an open fire. Flue-curing had not been developed to any extent in the 17th and 18 centuries. A good crop of tobacco yielded about 1,000 pounds an acre. One worker could handle from 1,000 to 3,000 pounds of tobacco, so a farmer without slaves could manage only one to three acres of tobacco.

Rice and indigo were both harvested by sickle. The indigo was cut as often as three times a year. Indigo required rotting and fermentation in vats, an eventual straining off of the liquid, and then a slow evaporation process until chunks of dye were produced which were cut into squares and boxed for shipment. The only change in indigo harvesting and manufacture during the 18th century was that around 1760 planters began using more oyster shells as lime in the fermentation process.

Rice harvest included cutting and shocking the grain, as with wheat. Next the rice was threshed by flail and winnowed by basket, and then the husks were removed, usually with hand mills. The remaining film on the rice was usually removed by pounding with pestles, although rice-beating machines appeared in the 18th century. After several siftings, the grain was ready to place in barrels for transport to market. For local trade involving little handling, the producer might put the rice in bags.

Harvesting in the North took longer than in the South because the shorter growing season with a longer fall and winter made it impossible to move immediately from a short harvest period into preparing for the next crop. The northerners did not have enough extra workers for a rapid processing of products. The ripe crops had to be promptly taken from the fields, but threshing and other processing operations could be performed later and the product marketed in small lots.

During the Revolution, farmers hedged against the inflation of money prices by disposing of their crops bit by bit at current prices. Until states passed laws regulating payment of debts and establishing legal values for currency at various times, debtors (which included most farmers) had an advantage during the periods of depreciating currency.

Everything had to be done in season, even if farmers could delay marketing. In New England, most threshing was done with flails on barn floors. The small farms, with small individual harvests and primitive modes of animal husbandry, made the flail preferable to treading the grain with animals. After the threshers had thoroughly beaten the grain a bundle or so at a time, they forked off the straw and swept the grain into piles which were shoveled into a bin and stored until winnowing.

Farmers threshed all grain in this way except for corn which was husked and shelled by hand. In the 18th century, very simple mechanical shellers were available. They usually missed the end kernels so someone still had to clean the cobs by hand.

Threshing inside the barn with flails may have been quite suitable for the uncertain weather of New England. Small amounts of grain were handled and marketed irregularly over most of the year. In the middle colonies, however, treading the bread and other small grains gained preference, particularly in the 18th century. For large crops marketed soon, farmers found treading by horses, or more rarely oxen, threshed larger amounts more rapidly than flailing and required fewer laborers. Using animals for treading worked less efficiently than flailing because they often failed to tread thoroughly, but the larger volume threshed reduced the economic loss per acre harvested. The lower labor costs apparently made up for the small amount of grain lost. A long harvest season with clement weather also encouraged treading which reduced the chance of loss through wet or stormy weather because farmers usually threshed outdoors on temporary wooden platforms.

The stacked grain would be brought from the field on sleds or in carts to the threshing floor. The farmer spread several bundles of grain in a circle and then walked the horses around it. One man stood in the middle holding the reins of the horses and guided them around on their path. When the grain had been knocked out of the straw, the harvesters forked the straw to one side, swept up the grain, stored or bagged it, and put more bundles on the floor. Sometimes boys rode the horses around in a circle. Two men and six horses could thresh around 100 bushels a day, or the yield of around six acres of wheat. For farmers with only two or three acres in small grains the method proved uneconomical. The system also depended on having horses. As the 18th century ended and the Revolution stimulated New England production, treading was used more in New England, especially for rye and barley. Consolidation of farms, with bigger acreages in grain, also encouraged a change to the less laborious treading method.

Various methods of winnowing had been devised to separate the chaff from the grain. The grain could be placed on a wooden or basket tray, tossed gently in the air, and caught again on the tray. The process continued until the grain was clean. Or the grain could be dropped from one tray to another with the chaff blowing off until gone. A gentle breeze aided the process. Otherwise workers used wicker fans or blankets to fan the chaff. Sometimes paddle fans, called

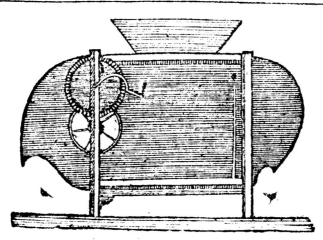

D U T C H F A N S,

F OR cleaning wheat or any other kind of grain, are made and fold by Adam Ekart, in Markét ſtreet, Philadelphia. Likewiſe rolling ſcreens, ſieves for ſifting iron ore, &c. warranted of the beſt make; alſo all forts of wire work, for cleaning wheat, barley, rye, flax feed, Indian corn, oats, or any other kind of grain, and wire ſhort-cloths for millers. The fame to be had of captain Matthew Phripp, in Norfolk.

Winnowing fan, ca. 1770s. (From *Virginia Gazette,* 1774)

fanning mills and powered by horses on treadmills, were used. This method of winnowing came into widespread use only after mechanical threshers appeared in the 19th century.

Chaff weighed less and was usually larger than the grain. Sometimes the farmers used a sieve, called a riddle, to sift the grain from the chaff. The riddle held the larger chaff and the very light chaff blew away as it and the grain fell toward the ground. The riddle was an important element in the mechanical thresher and winnower.

In all these processes the grower moved his grain with wooden shovels, scoops, and brooms or brushes. After winnowing, the grain could either be stored in wooden bins or put into bushel bags for handling and transport. The bags were usually made of light canvas or Osnaburg, a coarse linen. In the 17th and 18th centuries, grain was seldom moved in bulk. For overseas shipment it could be placed in barrels or shipped in bulk or in bags. A bag weighed about 60

Farm tools, *ca.* 1790. (From *Eighty Years of Progress of the United States,* Vol. 1, New York, 1861)

pounds when full. The bags had the disadvantage of concealing the contents. To check the quality of the grain, each bag had to opened.

Bulk transport of grain in carts or wagons could be undertaken at certain seasons for short distances to a mill, but the flour came back in barrels. Bulk transport across any distance was impossible because of the poor roads and lack of bridges. The vagaries of the weather also endangered bulk grain if the journey took over a day.

Two-wheeled carts provided most of the transportation for farm products, except in the middle colonies. The carts negotiated poor roads better than four wheelers. The Revolution stimulated road building by the several armies whose need for heavy transport increased the use of wagons throughout America except in the South. Braddock had built a road as he advanced, as had John Forbes in his attack on Fort Duquesne. Several of the roads built during the French and Indian War became routes for pioneer immigration. The changes in land transport came mostly from need and the creation of the capital to meet that need. In these matters the full importance of the Revolutionary War has yet to be uncovered.

Historians have traditionally emphasized the vast differences in agriculture as practiced in the northern, middle, and southern colonies during the 17th and 18th centuries. They were really not so great as suggested. Market improvements, related to the growth of American and European cities, gave rise simultaneously to the need for better implements and the capital to secure the more advanced technology. In 1775, Philadelphia had 40,000 people, a greater population than any English city except London. Other American cities provided increasingly profitable markets.

During the Revolution, cities grew only about 3 percent while the general population by natural increase and immigration rose by 40 percent, 1774–1790. The contending armies and the foreign markets compensated for this temporary slowdown in the domestic urban market. The war inflation allowed farmers to accumulate capital on a scale never before possible. Inflation sometimes allowed them to retire old debts with a depreciated currency.

The ravages of war consisted mostly of the loss of livestock and of growing or stored crops. Confiscation of estates merely changed the ownership of capital which did not diminish much. Farmers readily replenished crops and livestock though natural increase. The Revolution stimulated American industry because the army needed goods of all kinds. At the end of the war, a surplus of manufactured items remained and a greater industrial potential provided farmers with improved technology. The economic base created during the Revolution also allowed them to adjust rapidly to the growing needs of the European market.

CHAPTER FOUR

THE SEARCH FOR MARKETS ✖ 1607–1783

F ROM THE BEGINNING, Europeans exploited the continent for their benefit. The exploitation, however, required vast infusions of capital, both public and private. Entrepreneurs expected a profit from their ventures. Most Europeans who came to America in the 17th century expected to remain Europeans. Neither the settlers nor the capitalists expected commercial farming to become established. The promoters expected their colonists to become self-sufficient in food, but little more. Their main efforts would be to obtain commodities for trade in Europe.

The earliest English settlers in America sought gold, spices, furs, and timber. Gold and spices were never found in quantity. Furs, timber, and piracy were more successful, but not enough to support the colonists. The English founded other colonies for various reasons, but the three enterprises of furs, timber, and piracy (under the name of patriotism) always proved the most immediately profitable. Every new colony derived some income from these undertakings.

The Pilgrims of Plymouth paid off their debt to the merchant capitalists from the profits of the fur trade in twenty-one years. The Pilgrims and Puritans added cod fishing to the list of successful activities of Englishmen in America. The French and Dutch found the fur trade to be the most profitable enterprises from the beginning. The lower south of Carolina specialized in deer hides until well after the mid-18th century. The Pennsylvanians and Marylanders made their first profits in furs.

Supplies of food from Europe were irregular and inadequate. Shipments were most affected by conditions of war or peace in the Old World. Consequently the new Americans were quickly forced to feed themselves although the need did not seem obvious at first. In 1609, food shortages forced the governor of Virginia to order all settlers to plant a corn patch or face starvation. After some initial confusion, subsistence agriculture soon flourished.

BEGINNINGS OF COMMERCIAL AGRICULTURE
IN TOBACCO

Commercially profitable agriculture first developed in nonfood products. The first successful commercial farming began in Virginia, but the later history of farming in America has tended to obscure this early development. By ignoring the planters and farmers of Virginia, Maryland, and Carolina, historians have asserted that commercial specialization did not begin in America until the 19th century and thereby ignored two centuries of commercial farming! Commercial farmers generally misjudged the needs of the market and produced what Europeans neither wanted or needed. The Virginia farmers finally hit upon a saleable commodity—tobacco.

The native tobacco of Virginia did not suit the Europeans whose tastes favored tobacco from the Spanish possessions to the south. But John Rolfe found and successfully introduced a variety of tobacco in Virginia in 1612 which satisfied European tastes. The famous Orinoco variety from Venezuela was introduced and the tobacco industry of the Upper South began to flourish. By 1617, Rolfe had worked out a method for curing the tobacco so that it did not spoil in transit.

The settlers were so afflicted with "tobacco fever" that the Virginia governor, Thomas Dale, had to require farmers to plant two acres of corn per person in 1616! Tobacco sold at 3 shillings a pound, and at that price Virginia farmers could make fortunes. In 1627, they grew some 500,000 pounds of tobacco. Virginia and Maryland exported some 35 million pounds by 1700.

Under the pressure of heavy production, prices fluctuated greatly, especially after 1630; and prices for tobacco began to drop regularly around the 1660s. Then most farmers faced serious income losses in the tobacco regions and looked for ways of cutting costs. One way of staying in business was to increase production without increasing capital investment. Of the elements of production, the abundant land cost the least. The farmer could exploit it without even using fertilizer. Total production depended on the amount of labor the farmer could put in the field.

INTRODUCTION OF SLAVERY

The traditional labor source for the planter had been indentured servants, but this labor proved too expensive for most producers. The purchase price for servants was high, and they had to be released at the end of seven years or so and given their freedom dues. The former servants usually began farming in competition with the other farmers. To expand production, tobacco growers needed less expensive labor. So they introduced slavery.

Around 1674, after an agreement had been made with the Spanish, Virginia farmers began to use slaves imported from Africa. The combination of slave labor and falling prices drove many small farmers out of the tobacco business. Some others joined the ranks of small and medium planters. The inventories of landowners in Virginia and Maryland show the extent slavery came to dominate tidewater farming. In 1680, about 65 percent of the landowners in Virginia and Maryland had no slaves. By 1780, some 75 percent of the landowners owned slaves, and most of the slaveholders could not be considered small farmers. Tobacco remained the chief commercial specialty.

Throughout the 18th century the tobacco farmers of the Upper South accommodated the changing international market by exploiting land and labor, not by changing specialty. The overwhelming importance of tobacco was shown in the export figures for 1766 when Virginia and Maryland farmers sent 100 million pounds of tobacco to Europe. In that year, the value of the exported tobacco came to about £768,000 sterling out of a total of farm exports amounting to £880,000.

The impressive success of the tobacco farmers should not divert attention from other areas on the continent. Farmers in New England had difficulty in finding a truly profitable specialty. The New Englanders sought subsistence chiefly from agriculture. They turned to other enterprises for commercial profit. Most of these undertakings used farmers as their chief labor supply. Many of the industries operated on the farms.

BEGINNINGS OF COMMERCIAL FOOD GROWING

Subsistence farming supplied the food needs of the New Englanders and provided provisions for the several nautical enterprises. The farmers raised corn, rye, barley, oats, buckwheat, and peas. Wheat was successful as a bread grain until black stem rust appeared in 1660 and made wheat growing difficult and uncertain in most of New England. Massachusetts tried to revitalize wheat farming by offering a bounty in 1750, but no inducement could overcome the ravages of the rust. Rye and corn became the staple cereals, and "Rye and Injun," a bread made of rye flour and cornmeal, became the usual bread of New England. During the Revolution General Washington commented unfavorably on this black New England bread, but apparently it made a suitable bread for use at sea.

No commercial wheat production on any large scale took place from about 1660 to until well after the Revolution. Farmers tried to grow flax and hemp with little success except for local use. Barley declined to the advantage of Pennsylvania as New Englanders reduced their beer drinking and increased their cider consumption. Apple orchards flourished everywhere. By 1700 cider had become the most popular common drink.

New England farmers prospered as livestock raisers. Commercial animal husbandry began early in the 17th century, but it became important only about the mid-18th century. Farmers raised large numbers of animals by the time of the Revolution, and New England supplied a major portion of the cattle used by the Revolutionary armies. The big cattlemen appeared on the frontier of New Hampshire and western Massachusetts. The need for grazing land attracted many pioneers to the empty lands. New Englanders usually allowed all livestock—cattle, sheep, horses, and hogs—to graze in a semiwild state.

A significant trade in beef and pork, dried and salted, furnished provisions for the sailors of New England. Occasionally meats moved in international commerce, especially to the West Indies. Generally, exporters found costs of transport made it unprofitable to sell New England food on the world market. Livestock which could be driven to market offered the best returns because they furnished their own transportation. The casual methods of animal care resulted in sheep with dirty, torn fleeces and cattle and hogs with tough meat. New England had to import wool from England until the end of the colonial period.

The difficulties and costs of transportation, added to the fact the world food markets required few sources of supply, priced the New England farmer out of the world market. But the Yankee did have his own port cities and the provisioning business, and farming did pay for farmers near the coast. Only animals which walked to market could be profitable.

COMMERCIAL FARMING IN THE MIDDLE COLONIES

Commercial farming developed fairly rapidly among the farmers in New York, Pennsylvania, New Jersey, and Delaware. Most of the settlers had been farmers in Europe, and climate and soil helped too. But the rapid rise of commercial farming in this area stemmed mostly from the historical fortuitousness of developing last among the colonies. The comparatively late start meant that farmers elsewhere had already made most of the mistakes. But the experiences of others, although helpful, did not really explain the flourishing condition of agriculture in the middle colonies.

The farmers of the middle colonies were in a good position vis-à-vis the settlers of New England and the southern planters. The West Indies offered a good market. The towns of the North sometimes had to import food. Boston began to import as early as 1680. The planters of the West Indies, who concentrated on sugar and indigo, bought food from the mainland and even Europe.

The English took over New York in 1664 and Pennsylvania was settled after 1681. By then, farmers from New England and the South had already committed themselves to their crop specialties. Farmers in the middle colonies concentrated on wheat. The middle colonies became the major food producers and were known as the "Bread Colonies" by the 18th century.

The insignificance of black stem rust gave them an advantage over New England. Flour and wheat made up the major exports during the colonial period. In 1763, farmers of New York produced about 250,000 barrels of flour and bread for export, while Pennsylvania farmers grew some 350,000 barrels for export. The farmers of the middle colonies generally had larger farms than those of New England. In 1765, Chester County farms averaged around 135 acres, and Lancaster County farms had around 400 acres. The typical New England farmer seldom used over 40 acres, and his holdings had been considerably smaller in the 17th century.

The key to the prosperity was the foreign market. Philadelphia developed as a large market and important port and became a great shipping and meat-packing center during the 18th century. These functions gave rise to a local market of some significance. By 1750, some 7 or 8 thousand wagons supplied the city with produce, mostly vegetables and fruits.

Next to wheat, farmers found livestock to be most profitable, and they raised large numbers of cattle and hogs, and some sheep. Farmers of the middle colonies followed more advanced methods of animal husbandry than those in New England or the South. The better care given livestock probably resulted from the incentive provided by good markets. A range cattle industry developed in western Pennsylvania in the 18th century in response to a need for dried beef. Except for a few parts of Pennsylvania, the middle colonies suffered from a chronic shortage of dairy products during the 17th and 18th centuries.

The processing industries also prospered. Between 1763 and 1766, Pennsylvania and New York exported annually about £770,000 sterling worth of flour and biscuits. These exports were about the same value as the tobacco exports from Virginia and Maryland then.

AGRICULTURE IN THE SOUTHERN COLONIES

The Lower South, including southern North Carolina and all of South Carolina and Georgia, had its first significant influx of settlers in 1670 at Charlestown. The pioneers farmed to survive; corn was their major crop. Their main commercial enterprise was trade in furs and deer hides, a trade which dominated their economic life well into the 18th century. In 1731, the port of Charlestown handled 250,000 deer hides, the high point of that trade.

Around 1696, rice growing began in South Carolina. The commodity entered into the economy from the West Indies, and brought a new reliance on slavery. The rice planters found two profitable markets: Europe where the better grades sold well, and the West Indies where the common grades served as food for slaves. Other markets on the mainland developed, but accounted for little of the total production.

Colonial rice growers developed inland swamps and paddies, mostly man-made, and dryland rice also flourished. Rice soon outdistanced all other farm crops in the Lower South in value and in production. In 1770, rice exports from Georgia and the Carolinas came to some 150,000 barrels annually. By the Revolution they exported over 165,000 barrels a year. A barrel contained about 400 pounds of rice and sold in 1760 for 50 shillings currency, or about 7 shillings, 1 pence sterling per hundredweight. The price in 1770 ranged from 7 shillings, 6 pence to 8 shillings. Therefore, the crop of 1770 brought farmers and planters some £225,000 sterling (less than half the value of the tobacco crop of Virginia at the same time).

Some rice went to southern Europe, but the greatest share was sent to northern Europe and the West Indies. From 1730 to 1739, about 74 percent went to Germany, Scandinavia, and Holland; the West Indian trade absorbed about 15 percent of the total in 1768. By then northern Europeans received perhaps 6 percent less of the crop than they had in 1730–1739. The bulk of the crop went to northern Europe, usually by way of England; and the rice always sold for comparatively high prices. Right up to the Revolution rice planters experienced nearly continuous prosperity.

The second most important crop in the lower South was indigo, introduced experimentally as early as 1690. It received both British and colonial bounties from time to time. A long series of wars and tariff battles gave Americans an assured market in the steadily growing British textile industry. The bounties and the protected market made indigo extremely profitable for the planters who found themselves in the northernmost climatic range of the plant. Farmers and planters adjusted indigo to the northern climate in the 1730s, in response to bounties and to the closing of the market to Spanish and French indigo. Elizabeth Lucas, left in charge of her father's plantation in South Carolina, experimented with indigo and produced the first commercially successful crop in 1742. Planters took it up immediately and exported some 138,000 pounds from Charleston alone by 1747.

Indigo production merged nicely with rice production on many plantations. The planter could use land for indigo which did not suit rice, and the seasonal labor needs of both differed enough so slaves could be given year-round employment. Indigo plantations also made fairly efficient use of slave labor, and most indigo was grown by planters who specialized in it. Indigo may have been responsible for the increased importation of slaves in the Lower South during the 18th century.

The bounty on indigo did not end effectively until 1783, but by then the planters had lost their market. During the Revolution other producers filled the American void in the British market. South America and India had a more suitable climate for the crop than North America. Consequently the loss of preferential treatment for American indigo caused farmers to diminish their planting. About 1787, the indigo planters had already shifted to the production of cotton. After the invention of the cotton gin, cotton completely replaced indigo. Simultaneously cotton culture spread rapidly into the interior of the continent.

WEST INDIAN TRADE AND SLAVERY

In general, the southern economy of the 17th and 18th centuries depended on trade with Europe and the West Indies. Southern farmers and planters could not possibly dispose of their surplus on the American market. For different reasons, both the Upper and Lower South used slave labor in large quantities. In the Upper South, the impulse was spurred by a need to reduce labor costs. In the Lower South, slavery grew largely as an effort of farmers and planters to stay away from the swamps and paddies because of malaria and yellow fever. No one knew what caused these diseases, but they were associated with stagnant water. White planters developed a convenient theory of African slave immunity to the diseases, and perhaps blacks did resist malaria fairly well. Working with the highly profitable rice and indigo crops, slaves returned large profits for most planters.

The greater importance of the European market should not obscure the substantial market in the West Indies. By the late 17th century, the West Indian settlers were almost exclusively producing sugar for the sugar-craving Europeans. This development took place throughout the islands: English, French, Spanish, and Dutch. Various imperial trade policies fostered the industry as each European nation sought to secure its own supply of sugar and to sell to those Europeans without American colonies. The West Indian farmers and planters concentrated so heavily on sugar that they had to import almost everything else by the 18th century.

The West Indian planters needed wood for buildings, houses, and barrels. They provided an insatiable maw for barrels and barrel staves. Everything from sugar to molasses to rum was shipped in barrels. Often ships carried coopers on board to make barrels while en route to the West Indies. Just before the Revolution, the islands took at least 11 million staves and headings and some 36 million feet of boards a year. Yankee farmers made most of the staves during the winter.

The planters of the West Indies depended on imported food. The ratio between commodities generally remained fairly stable. Thus one acre of sugar would produce enough revenue to buy the yield of five acres of corn. Other commodities had ratios favorable to the sugar planter, and so planters preferred to buy most of their food and use their own land for sugar. Corn, wheat, flour, and bread from North America always found a ready market in the West Indies. The West Indians also bought large quantities of dried and salted meats. In 1770, the total value of all American exports to the British West Indies came to £844,000 sterling. (The value of all exports to Great Britain came to £1,636,000 sterling.) If the unofficial illegal trade with the foreign West Indies had been included, the West Indian trade would have appeared to be even more important for the Americans. As it was, the West Indian trade came second only to the trade with Great Britain.

The high proportion of food exports to the West Indies had considerable influence on the total development of American agriculture, especially on the Bread Colonies and probably also the Lower South. Americans could readily reach the West Indies by sea. Even a domestic market in America would have had limited influence because of the difficulty of internal transportation. Grains and meats could not have found a large ready market in Europe. In the absence of the island consumers, American farmers might have been stuck in subsistence doldrums for another century. Or until American cities grew, and Europe became markedly food deficient. The rise of commercial agriculture early in American history provided the capital for other enterprises.

SLAVE TRADE WITH AFRICA

The trade between American farmers and West Indian planters led to more trade between America and Africa. Beginning around 1674, the trade centered chiefly on slaves from Africa and by several indirect routes involved American agricultural produce. The Royal African Company of Britain received a monopoly on the trade between Africa and the American colonies in 1672 and introduced slaves in large numbers beginning in 1674. By way of the West Indies, American products were exchanged for products wanted in Africa. The Royal African Company made low profits and had incompetent management. Having begun slavery on the mainland it failed to satisfy a growing demand for slaves. The company lost its monopoly in 1697, and thereafter the trade flourished as both British and American slavers entered greedily into the business. The number of involuntary emigrants to the Americas from Africa has been variously estimated, but the figure must have reached 20 million between 1519 and 1808.

The British forced the emigration of some 5 million Africans to British colonies in America. Less than a tenth of these came to North America. The continental colonies had about 224,000 whites and 28,000 blacks in 1700. The mainland had around 400,000 slaves by 1760 and about 2,100,000 whites and some 500,000 blacks in 1774. The census of 1790 included around 698,000 blacks. Only a few of these people of African descent were born in America.

In the 18th century, New England shipowners were leaders in the slave trade. The colonials used small ships, "slavers," which usually ranged from 50 to 60 tons and carried a crew of about half a dozen. The British were involved on a larger scale both in size of ships and number of ships. In 1771, out of about 265 British and colonial ships engaged in the trade, the Americans had about 70, mostly from New England and New York.

The slave trade involved the famous triangular trade. Typically the slaver carried lumber, fish, breadsutffs, and some gold, and these were traded for rum in the West Indies. The ship then sailed to Africa and the rum and gold were traded for slaves. In 1750, a male slave cost about 100 gallons of rum, a female about 85 gallons, and a child around 65 gallons. (The trader invariably watered the rum and gave short measure whenever possible.) The slave freight, around 80 persons, would then be brought to the West Indies and sold for an average of £20 sterling each. Prices varied by sex and age. Planters or middlemen paid with molasses and gold. The molasses, in turn, usually went to New England to be distilled into rum which entered into the Indian fur trade and also into the African slave trade.

The amount of rum diverted to each enterprise has not yet been reliably estimated, but most of the New England rum probably went into the Indian trade. Thus the debauchery of Indians and the enslavement of Africans arose from the distilleries of New England. Some slavers went directly from New England to Africa, and many slaves ended up on farms in Massachusetts, Connecticut, Rhode Island, New York, New Jersey, and Pennsylvania. These colonies later claimed they did not use many slaves, and most of the African immigrants did go to the southern colonies. The northerners did follow a more generous policy of emancipation so that by the time of the Revolution places such as Cambridge and Roxbury already had sizeable populations of northern-freed blacks.

Southerners also received direct shipments from Africa, although some planters preferred to obtain slaves who had been worked some time in the West Indies and thus acclimated. The African culture of the slaves was not completely destroyed. General William Moultrie, returning to his plantation in the midst of the Revolution, was welcomed by his slaves with an African war dance and an extended feast celebrating the return of the warrior. Although Americans usually bought slaves who had spent some time in the West Indies, in retrospect the theory of acclimation seems based on a fictional concept of the usefulness of a sojourn in the islands. Perhaps slave dealers in the West Indies fostered the idea.

Slave importations to the continental colonies averaged 2,500 a year from 1715 to 1750 and 7,500 a year from 1760 to 1770. Taking a rough average of £20 sterling for each slave, the business grossed around £146,000 a year after 1760. The slave trade did not amount to much in commercial terms compared to tobacco or furs, but it had a tremendous impact in terms of immigration. It also set the plantation pattern of southern agriculture to a considerable extent, although probably the nature of the commodities had more influence than the organization of the labor system.

INTERCOLONIAL TRADE

Although intercolonial trade was great, it amounted to far less in value than foreign commerce. However, intercolonial trade was greater in volume. Most commodities moved by water. Few settlements of the 17th or 18th century were located far inland until after the Revolution. Traders could reach most places of any size by river which led to a proliferation of small sailing vessels and to the development of the equivalent of keel boats—barges pulled from shore. Farmers and traders used a variety of canoes. New Englanders and later New Yorkers and Pennsylvanians built and operated ships expressly for the American carrying trade. At the close of the fishing season, New Englanders often loaded their fishing ships with all manner of agricultural products including beer and entered the coastal trade.

In the early 17th century, inland transport used Indian and animal trails suitable only for horses and pack animals. By the end of the century, governments and individuals sometimes widened the trails enough to allow the use of some two-wheeled carts and four-wheeled wagons. A road connected Boston with Hadley on the Connecticut River by 1674. Roads also linked other points in New England and Virginia. By 1716, one of the earliest coach lines in America began running between Newport, Rhode Island, and points in Massachusetts. Chester County farmers in Pennsylvania could ship by road to Philadelphia and then to the great world markets by 1710.

Road building in the 18th century still consisted mainly of widening trails and paths with large stumps left undisturbed. Cart and wagon wheels had to be large to clear these obstructions in the roads. Fords and occasional ferries gave passage across streams, and now and then a bridge appeared. American roads, in spite of inadequacies, compared favorably with those of France or England by 1700.

From the 17th to the 19th centuries, cattle droves were much of the traffic. Cattle drovers trailed cattle from South Carolina to Philadelphia and New York as early as 1740. Drovers from all over New England had moved herds of cattle to the towns by the mid-17th century. Accommodations of sorts for the animals were available, and inns and taverns provided stopping points for the drovers. Corn-fed cattle and hogs which transported themselves provided one important way of moving the corn crop to market. The grain otherwise had to be transported by cart and wagon.

In the middle colonies appeared the Conestoga wagon, the great freight wagon built something like a boat for fording streams and covered with canvas to protect the cargo from the weather. Built by German craftsmen in Pennsylvania, these wagons soon appeared nearly everywhere in rural America. By 1750, Americans had at least 7,000 Conestoga wagons, owned chiefly by traders and inland merchants.

The peddler and his cart had appeared early in the 18th century and continued his business into the 19th century. Throughout the colonial period, pack animals continued in use for many products, including alcoholic beverages. Carts provided the chief transport for farm vegetables and small grains. (The grain was put in sacks or barrels.) Wagons of various sorts were used much more as the 18th century ended. For example, General Edward Braddock easily found supply wagons. If he had had fewer wagons, his troops might have maneuvered more easily at the Battle of Turtle Creek.

EFFECT OF TRANSPORTATION ON COMMERCE

Poor roads and the great distances limited what a farmer could profitably sell and where he could sell it. With expensive, infrequent, and slow transport, trade was largely confined to the least perishable farm commodities. Farmers in New York and Pennsylvania sent grain and flour to New England, the Carolinas, Georgia, and the West Indies. They sent beer to the South and to Canada, and homemade woolens and wool all over. Naval stores, such as turpentine and pitch, went to coastal towns. The comparatively imperishable tobacco and rice went to the North, and indigo and sugar were added in the 18th century.

Commerce directed the nature of the agriculture. Perishability, the distance to a market, and the value of the product in the metropolitan or European market determined farmers' specialties. The considerable distances, the slight demand for many products, and the slowness of the transportation also limited the types of commercial products. Climate played a role; tobacco grew better in the South and wheat in the North.

Farmer fairs and market days supplied the cities with meat and other agricultural produce. Merchants and country people even held fairs in the vicinity of army camps during the Revolution. Despite commercial specialization, most farmers also had to feed themselves and did some general farming. Southern farmers engaged heavily in commercial farming and relied on the northern colonies for much of their food. However, the subsistence farmers were the poorest, as those in the Virginia backwoods where subsistence farming predominated at the end of the 18th century.

The fastest and cheapest transport was by sea or river, and so commerce tended to be coastal or with Europe. In the late 18th century, it cost considerably less to ship from England to Philadelphia than to ship iron to Philadelphia by land from Lancaster only 75 miles away. Freight could be sent more cheaply from Pittsburgh down the Ohio and Mississippi and then by sea to Philadelphia than overland from Pittsburgh to Philadelphia. The trip to England from Philadelphia cost less than any route between Pittsburgh and Philadelphia.

Grain shipped from western Massachusetts to Boston in the 18th century moved short distances by land to the Connecticut River and by water to Boston. The greater part of the trip was by water, yet the land part of the voyage cost $2\frac{1}{2}$ times more. The cost of land transport depended on the season of the year. The wet spring made roads more difficult and the cost rose. Summer and fall transport cost less. In the late 18th century, land transport from the back country of South Carolina to Charleston cost about twice as much as the transport from the Connecticut valley to Boston. The wagon rate from the Shenandoah valley in Virginia to Philadelphia cost four times as much as a carriage from upper Carolina to Charleston. In short, a commodity had to sell for high prices to pay its costs of transport and handling and still leave anything for the farmer.

Ocean transport was neither swift nor cheap. Ships had small capacities throughout the colonial period. In the 17th century, a typical ocean ship had about 120 tons capacity, but by the end of the 18th century such ships carried around 170 tons. (In the 1960s such ships carried around 9,000 tons.) Some improvements in shipbuilding made the trip of the 18th century a little faster than that of the 17th century. However, the voyage from England to America took from 50 to 100 days or more. The return trip from America to Europe took only 15 to 25 days because of the prevailing winds. Ships rarely managed more than two round trips each year. However, this European trade surpassed all other in value, and from this trade sprang the main prosperity of the American farmer.

CHAPTER FIVE

MERCANTILISM AND THE FARMER ✠ 1607–1783

W ITH THE WANING of the Middle Ages and the rise of nation states, European businessmen and statesmen began to think in terms of national goals as well as private profit. The concepts of mercantilism reflected a broader view of life than the localism inherent in feudalistic concepts. A strong nation provided businessmen with good profit-making opportunities. The early economists theorized that a government should run the national economy by the same principles as people ran their households. Among national objectives, the accumulation of gold within the national borders ranked first. Most men felt gold represented true wealth. To accumulate gold, the businessmen of the nation had to sell more than they bought or gold would leave the country.

Statesmen could more nearly realize the ideal of mercantilism if the nation achieved economic self-sufficiency. This led, in turn, to the development of colonialism. What a country could not produce itself, it could secure from colonists and avoid buying on the world market. The apparent success of the Spanish in America and the Portuguese in Asia strengthened the conviction that the system could work. Colonies came to be valued as auxiliaries to the mother country. In North America mercantilist practices seldom succeeded before becoming distasteful to the inhabitants. In the 16th century the Dutch prospered as a nation without following mercantilist theory, although the Dutch bought more than they sold. Their empire was small and their success apparently did not depend on colonial exploitation. The Dutch made up their deficit trade balances by adding shipping charges and insurance premiums.

Soon other countries added another element to mercantilist practice that the Spanish had long followed—the requirement that all commerce between the mother country and her colony move only in ships of the mother country. Such a regulation curbed foreign competition, and the nation found another source of profit in the carrying trade. Acting on this idea, the English successfully suppressed the Dutch carrying trade to America. They also treated American ships as British, and the colonists quickly seized a disproportionate share of the carrying trade, which irritated the English. American farmers and planters mostly benefitted from mercantilism but ultimately not enough to want to stay in the empire.

NAVIGATION ACTS

In 1620, the Crown and Parliament ordered all Virginia tobacco carried to England before being sent to the rest of Europe. In 1624, England required all tobacco to be carried in English (also American) ships. These acts established the principles for a long list of navigation acts. They also secured several advantages for the tobacco industry, including a prohibition against the English farmers growing tobacco. The Virginia monopoly with an assured market and the development of regular commercial connections benefitted the American farmers and none protested the acts. The apparent success of these laws encouraged the application of more regulations. The Crown issued orders designed to put the American import trade in the hands of Englishmen and Americans, but laws regulating trade in America tended to establish precedents rather than loyal obedience.

The Navigation Act of 1651, passed by the Puritan Parliament during the Protectorate, marked the first significant effort to impose an imperial mercantile policy on Americans. The act required that all goods from Asia, Africa, and America going to Britain had to go in British or American ships. The act, intended primarily to cut the Dutch out of the carrying trade, also required the use of British or American ships in trade with Europe. Only ships of the country which produced a commodity could be excepted from the act. Thus a French ship might bring French wines to Virginia, but it could not bring in Spanish raisins. The act assured American shipowners of a share in the colonial commerce. The modest stimulation of colonial commerce probably helped farmers. Americans, thus assured of cargoes, could venture into the risky business on a larger scale and provide more and ultimately cheaper shipping for the farmers.

In 1660, the Crown and Parliament strengthened and broadened the earlier acts. The Act of 1660 reasserted the rules of the Act of 1651 which not only proclaimed the return of the Stuarts and the end of Cromwell, but also suggested that the Act of 1651 had not been closely observed. The Act of 1660 enumerated certain commodities deemed essential for British prosperity and safety which could be exported only to England. The importation of commodities from anywhere except Britain or her colonies was prohibited. The British thus forced the American farmers and planters to use only one market, but the colonials had a monopoly within that market.

The benefits of monopoly almost always outweighed the disadvantages of shipping only to England which as the sole customer had the best machinery for reexport. The enumerated items that received special favors included sugar, tobacco, indigo, dyewoods, and ginger. Probably at the insistence of planters and miners, Parliament later added rice, molasses, naval stores, furs, and copper ore. The colonists naturally approved the regulations which amounted to a subsidy because they excluded foreigners from the British market.

In 1663, the Parliament passed the Act for the Encouragement of Trade, usually called the Staple Act. This act supplemented the Act of 1660 by regulating foreign goods going to America. Foreigners had to land their goods first in Britain and pay a duty before shipping to the colonies. Rum from the French West Indies would first be shipped to England and then back to America. Rum from the British West Indies, however, could go directly to New York. If enforced, the act obviously protected the colonial market for both colonials and Britishers.

In addition to granting American farmers a virtual monopoly within an assured market, the Act of 1660 had also imposed small tariff duties on the enumerated products. The tariff did not seriously reduce the prices paid to American farmers. Since British consumers had no source for the enumerated items other than America, they paid the taxes and bought the products. But additional profits could be made by smuggling, and the consumer paid less for smuggled commodities. Smuggling flourished in England and particularly in Scotland.

To restrain smugglers in the mother country, Parliament in 1673 required that the duties on enumerated commodities be paid in the colony of origin. This "plantation duty," created to halt evasion of the earlier acts, cut down smuggling. No one could profit much from smuggling something that had already been taxed. The plan worked fairly well, suggesting that smuggling centered in Britain rather than in America. The significance of the act for Britain and America showed up in the increase in revenue. In 1685, the Crown derived from £100,000 to £130,000 sterling from the tariff on tobacco alone, the most profitable of the enumerated commodities.

EFFECTS OF TARIFFS ON
AMERICAN TRADE AND SMUGGLING

The tariff acts allowed the shipper to receive a rebate of one-half the duty if he reexported. Eventually those who reexported sugar and tobacco received all of the tariff back. The tariff soon ceased to concern the American farmers and planters, whose products mostly went duty free. Holland, Germany, Ireland, Sweden, and Russia provided large markets for Britain throughout the 17th and 18th centuries, particularly after 1700. By the time of the Revolution, 80 percent of the exports of Virginia and Maryland went ultimately to the Continent. Some went directly, but the bulk moved through regular channels of the protected market which had many attractions.

On the other hand, the Americans were affected by the prohibition on direct exports from foreign nations and colonies. The Spanish, French, and Dutch West Indies supplied important amounts of rum, sugar, molasses, and slaves. A large smuggling trade developed in slaves and sugar. In 1696, Parliament attempted to control this illegal trade by providing that violations of the Navigation Acts, especially that of 1663, should be tried in admiralty courts. The government sought to avoid regular trials in America because of the leniency of the jurors. Americans defied the Act of 1663 and smuggled rum, sugar, and molasses. Parliament attacked foreign sugar and molasses with heavy duties in 1733 and ordered an honest effort at enforcement of the law in 1763.

The Americans ignored both laws because to have paid attention would have disrupted the rum-furs-slaves-rum complex. The slave trade and the Indian fur trade, both highly important and related to agriculture, would have been severely hampered if trade with the foreign West Indies had been seriously restricted.

American farmers did well during the 17th and 18th centuries—especially during the first three-quarters of the 18th. But they seldom counted their wealth in hard cash. For a variety of reasons, specie (money and bullion) continually flowed from America to England because Americans usually bought more than they sold. Prosperity on the farm and in other sectors of the economy required a steady source of money. No source of money produced so well as the Spanish Main, and no agents so effectively brought it out as pirates and smugglers. Because of the need for coins, colonials almost continuously violated the Molasses Act. American producers sold to smugglers as well as to ordinary shippers because, outside of the South, most agricultural products brought the best returns in the West Indies.

Piracy had been an important economic activity from the beginning of the colonies. Setting up bases for piracy was one of the reasons for founding Virginia, Plymouth, and Massachusetts Bay. However, the adventurers stated their objectives in patriotic terms of defense. Pirates apparently reached their peak numbers and importance between 1690 and 1720, their "golden age." They operated chiefly in the West Indies and on the Spanish Main. Pirates normally seized the commerce of nations other than their own; and in a time of nearly incessant warfare, they often managed to combine patriotism, profit, and religious prejudice in one glorious cause.

SUPPRESSION OF PIRACY

American colonial officials often gave passes of safe conduct to pirates and took bribes to give pirates safe haven in American ports. The number of persons involved and the profits were both high, although because of the nature of the business, incomplete records were kept! Some figures are known. In 1699, Lord Bellomont reported forty or more pirates in custody in New York, Connecticut, and Pennsylvania. This number probably was that small handful unfortunate enough to be apprehended. One ship cleared some £30,000 sterling on one voyage. The Phillipses of New York were charged with having made more than £100,000 in the business. Suppression of piracy began seriously around 1720, and the British navy had ended most of it by 1750 with almost immediate agrarian repercussions.

The suppression of piracy helped cause a shortage of currency. Pirates normally spent their profits instead of burying them! One authority believed that in 1750 over half the coinage in circulation in America came by way of piracy. Even if the figure were inflated, a high percentage of money was brought in by pirates. The end of this money flow made trade with the foreign West Indies even more essential and thus made the Navigation Acts more objectionable. Smuggling was the recourse for Americans.

Many American merchants and shipowners simply ignored the acts. They felt the laws had been passed for the benefit of British merchants, but the acts actually seem to have been passed mostly for the benefit of colonial farmers and planters. However, by preventing trade with the Spanish, French, and Dutch West Indies, the acts hurt colonial shippers and also cut down on the possible markets for farm produce. Still, Americans carried on the illegal West Indian trade, and they also traded directly and illegally with France and Holland. Until after the mid-18th century, port collectors received their pay from the colonial assemblies Consequently these British officials often yielded when the colonial assemblies sometimes favored the smugglers. Landing goods on the coast of England without paying duties presented no problem to skilled English smugglers. In 1700, possibly as much as one-third of Boston's trade violated one law or another, especially those regarding the West Indies. In England at the same time, as many as 40,000 smugglers carried on their profession.

Smugglers carried goods between the South and the West Indies and between the South and Europe, but most smuggled out of New

England and the middle colonies. Foreign goods came from Europe and the Far East. Until 1763, British officials usually did not seriously try to stop smuggling. When the British did enforce the laws, trouble, especially agrarian, followed. Enforcement cut off the money supply already made short by the suppression of piracy. Commercial farmers had to have money.

Enforcement first involved the use of writs of assistance to find smuggled goods A Crown officer would issue an indefinite search warrant specifying neither the goods sought nor the place to be searched. Beginning around 1761, James Otis, lawyer for the smugglers, and John Hancock, smuggler, especially opposed these writs of assistance. The most serious blow to smugglers came in 1764 when Parliament passed the Sugar Act which lowered rates on sugar and molasses and set up machinery to collect these duties. The act provided for patrols to catch the smugglers and required shipmasters to post heavy bonds to assure observance of the laws. The law increased duties on many other commodities and raised the number of enumerated goods, but smuggling continued despite these disabilities.

COMMERCE DURING THE REVOLUTION

During the War of the Revolution, agricultural marketing suffered some changes and many disruptions. Foreign commerce continued to form the basis of most agricultural success. Regular commerce continued although embargoes and other devices of the Congress sometimes severely restricted or curtailed trade. The chief commercial activity centered on privateering, or legalized piracy. Some privateers were especially built for the business and could outfight the ships they could not outrun. Most privateers were just heavily armed merchantmen trading with France and the West Indies. They captured enemy shipping when opportunity offered.

The Americans completely ignored British regulations and had little difficulty securing trade rights in the French colonies. Trade with the Spanish colonies, although extra-legal during most of the Revolution, took place with official Spanish blessing for the most part. Americans supplied the West Indies with tobacco, wheat, and rice. The Dutch, also allies, furnished the first secure trading post on the island of St. Eustatius. The British captured the island in 1781, whereupon trade was transferred to the Danish islands of St. Thomas and St Croix. From these posts the goods were then sent to the other islands in neutral ships. Some of the goods, such as indigo and tobacco, went to Europe under various flags. In return, the islands formed a depot for arms and ammunition for the Americans. The French and Dutch entered heavily into the trade and the Swedes and Danes also took part in the commerce on a lesser scale. The American farmers, if unmolested by British or Tories, did enough business to survive and some even became rich.

Internally, the market patterns shifted markedly, chiefly because the contending but moving armies formed the largest domestic markets. The commissaries of both sides directed farm produce to the troops in the field. Soldiers in the American armies sometimes had enough food to send some home to families caught short. The growth of American cities nearly ceased, and they may even have lost population in the early days of the Revolution. Although Philadelphia was larger than any English city except London in 1775, the urban market now grew little. Occupying English forces in the towns provided a market and usually paid for their food. They had to import a great deal of it because American farmers preferred not to sell to them.

Farmers far from the armies usually felt little economic impact from the war except for taxes and depreciating money. Much of the rolling stock of wagons and carts, as well as horses, were absorbed by the armies and were not easily replaced. Only livestock drovers could move their commodities with any ease and certainty. In the southern plantation states, farmers needed more self-sufficiency than before the war. They especially suffered from destruction of farms and crops by the contending armies after 1779.

Enough food was produced throughout the Revolution to feed the armies, maintain the civilian population, and even produce an exportable surplus to exchange for war supplies. Food shortages in the American army largely reflected shortages of transportation. Congress sometimes had to prohibit the export of certain foods from certain ports. Destruction of food supplies by the various armies caused local food shortages, especially during British occupation of an area. Although sea shipment of food was risky, such shipments generally succeeded. The British navy sometimes closed various port briefly. Occupation by Redcoats worked better.

New England coastal towns such as Boston and Salem, had come to depend on the middle colonies for much of their food by the time of the Revolution. During the war, New England always got food. New England farmers and consumers alike had to overcome the difficulties of overland transportation. The farmers did produce enough. Shortages resulted from other difficulties in a war-torn country. Most farmers favored the American cause and sent their food to the American armies. The British could command farm produce only as far as their patrols could reach to seize it.

States sometimes had to prohibit the distillation of grain or order the channeling of food to the armies away from more profitable foreign commerce. Congress levied taxes of commodities, especially food. The years 1777 and 1778 were bad crop years. Bad weather caused local shortages and famines. Some states put embargoes on driving livestock out of the state, particularly when it was invaded.

Americans had overcome the difficulties of handling commodities in new routes of trade. Country stores and small merchants in villages suddenly had to handle large volumes of food with inadequate facilities. Many larger merchants chose the Loyalist side so their knowledge and contacts were not available, but their facilities were often confiscated. Impressment or confiscation of food, paid for by national and state certificates or quartermaster receipts, became a common way of raising food for the army. This method became more necessary as all currency, particularly the Continental, steadily declined in value.

The British had a constant stream of supply ships running between Europe and America, carrying food for the troops, fodder for their horses, and even winter fuel. Except for rum, Americans never needed to import food. The record does not show a single American commander who had to significantly alter a tactical plan because of a food shortage. American generals sometimes fought ill-advised actions, such as when their troop enlistments were about to run out, but never because of a food shortage. The British often had to alter lines of march to fit their food needs. Despite occasional shortages such as those at Valley Forge, the American farmers kept their armies fed. The British brought their food from overseas or sometimes grew it themselves during a long occupation.

The threat to commercial farming implicit in late 18th-century imperial trade policy had strongly influenced the drive toward American independence. Professional men and merchants led the debate, but American farmers did the fighting. Conspicuously, a Virginia planter bet all he had on a successful outcome and then as Commander-in-Chief fought to win. Farmers fired "the shot heard round the world." They made up most of the population, and as commercial farmers they also had much to gain.

The reality of American farm prosperity appeared obvious by the end of the colonial period. During the Revolution, German, French, and British officers often commented on the high standard of living of the American "peasant." These observers left their impressions in writing, but their peasant troops made their evaluation with their feet. High levels of desertion of Irish peasants from the Royal Army and the joining of these peasants and of German mercenaries to the ranks of American farmers suggested the appeal of American farm life.

CHAPTER SIX

LAND LAWS AND POLICY ※ 1785–1865

THE ASTOUNDING RISE in total population, the slow increase in urbanization, and the increase in agricultural production all resulted from an abundance of rich American farmland. It was time to enact laws and policies that would allow the rapid transfer of the land to private individuals. By 1788, the national government under the Constitution had taken shape. Nearly all the states had given up their more extravagant claims to the interior, but they still maintained their own land systems within their own political jurisdictions. The bulk of American land had come under the control of the Federal Government.

The operative land disposal system was that laid out in the Land Ordinance of 1785. Congress had modified some of the more obvious inequalities when it set up land offices in several places, generally nearer to actual settlement. But farmers still had to purchase land in large lots without any credit backing from the central government. As a result farmers protected their interests with claims associations which either bought land for them or, more often, forced equitable arrangements from the land speculators who bought the land from the government. Representatives from the claims associations were always impressively armed, and they tended to intimidate the agents of landowners. As for squatters, the few government officials could seldom dislodge them. George Washington was one of the few landowners and speculators who had any noticeable success at moving squatters off his land. In confirming his claims he had the prestige of a "Hero of the Revolution." Most speculators had less success.

The first Federalist administration intended to use the land primarily as a source of revenue, an objective which continued the policy of the previous Congress of the Confederation. The Congress of the Constitution had broad taxing powers which ensured adequate revenue without selling land. In 1790–1791, Alexander Hamilton proposed as part of his general fiscal program that the survey and disposal system of the Act of 1785 be abandoned in favor of indiscriminate location and survey by metes and bounds. Other revenue-producing aspects of the law would remain in force. He intended to speed the sale of land, reduce the cost of surveying, and produce more revenue in less time. In general, Hamilton's suggestions favored speculators over farmers. Congress did not act on his proposals.

ACT OF 1796

Frontier farmers had virtually no representation in Congress until new states began entering the Union. Tennessee entered in 1796 and Ohio in 1802. The popular pressure for a change in the land laws began to increase. On May 18, 1796, Congress had passed a new land act, the first since 1785, which continued the survey system, but made some fairly significant changes, all designed to produce revenue for the Federal Government. The law raised the price per acre to a $2 minimum. Bidding on land began at that price, but the law granted the purchaser a small amount of credit. A year might elapse between purchase and payment.

The act reduced the size of the land parcel. Alternate townships were no longer sold entire, but instead in chunks of no less than eight sections. The other townships were to be sold in sections. A section of 640 acres still far exceeded what a farmer could actually farm. The law also set up land offices nearer the land being sold, with offices at Pittsburgh and Cincinnati. A Surveyor General and a corps of surveyors were established, and a neophyte bureaucracy thus emerged to deal with land problems. The system did not cause any increased sale of land. Only 48,566 acres had been sold by 1800. The Federalist administration was by then willing to entertain some changes in the law to increase sales and revenue.

ACT OF 1800

As a result Congress passed another land act on May 10, 1800. William Henry Harrison presented the act which he intended partly to stimulate sales and partly to continue protecting the interests of speculators. The act continued the minimum price of $2 an acre, but permitted four years of credit. Purchasers had to pay one-fourth of the price in forty days, another fourth in two years, and the remainder by the end of the fourth year. These time payments carried a charge of 6 percent a year. Those who paid cash received a discount of 8 percent. The speculator could make the down payment on a large amount of land, hoping to sell enough small parcels to make subsequent payments. The act cut the minimum purchase to 320 acres, but this was still more than the average farmer needed and allowed another level of speculation. In Ohio, land offices were set up at Cincinnati, Chillicothe, Marietta, and Steubenville.

The act obviously stimulated land sales. By November 1, 1801, the government had sold some 398,466 acres. Sold; but not all paid for. The pioneer farmers wanted a preemption act under which they could buy land they had improved at the minimum price. Congress rejected this proposal. Farmers in states with public lands still needed more representation in Congress. When the Jefferson administration took office in 1801, the land laws were in for serious revision.

When Ohio entered the Union, Congress voted to keep all ungranted lands, except one section in each township which was given to the state to support education. This policy continued so that the Federal Government became the largest single landowner in many states in their early years. The amount of land granted to the states for education gradually increased. Some far western states ultimately received four or more sections in each township.

Continuing pressure by farmers and their representatives, plus the advent of an agrarian-oriented administration, led to further liberalization of the land laws. On March 26, 1804, Congress reduced the minimum bidding price to $1.64 an acre and the minimum acreage to 160 acres. This acreage still exceeded what the typical farmer could farm. The credit provisions of 1802 continued in force, requiring only a one-fourth down payment. Viewed in another way, the purchaser only put down total payment for 40 acres, which came close to what an average farmer could actually use.

INCREASE OF LAND SPECULATION AND THE LAND LAW OF 1820

The credit provisions and the lower minimum acreage encouraged speculation on a vast scale because the risk of loss was slight. Only half the land sold by the United States had actually been paid for by farmers or speculators by 1819. When farmers and others found they could not meet the payments, they asked Congress to allow them to keep the land already purchased outright. By 1820, Congress had passed twelve such relief acts for purchasers. Default had become nearly institutionalized. Congress in effect underwrote land speculation by anyone who could make a down payment on 160 acres. However, land was still sold at auction, and most land sold well above the current minimum price. The land speculation caused widespread overextension of credit and inflation which were partly responsible for the Panic of 1819 and the depression of 1819–1821. Congress passed the Land Law of 1820 to cut down on the land speculation, which was the chief cause of the nation's financial difficulties.

The extent of the speculation had reached extreme proportions. The Federal Government had sold some 19,339,158 acres from 1800 to 1820. Because of reversion to the government for nonpayment, only 13,649,641 acres had been actually transferred. In 1820, purchasers still owed the Federal Government over $21 million for land. Many comparatively small farmers had defaulted, but southern planters had also expanded their holdings. The most land speculation took place in Alabama and Mississippi in response to the great boom in cotton growing. By 1820, one-half the total owed to the United States for land was due from farmers and planters in Alabama and Mississippi alone.

The Land Law of 1820 abolished all credit for land purchase, but allowed the sale of as little as 80 acres. The act recognized the agricultural needs of the farmers. The act also cut the minimum price to $1.25 an acre. Congress again refused to enact preemption to protect the actual farmers. Pioneer farmers still had to form claims associations to protect their interests at land-office public auctions. Western voters and congressmen did not unanimously support the cash payment provisions of the law which passed on April 2, 1820.

Because farmers did not have ready cash, the law still favored speculators. Land sales fell sharply after 1820, with only about 3,000 acres sold in that year. This low sales record reflected the general depression of 1819–1821. Sales rose after 1823, and a total of 56,501,731 acres had been sold for cash by the end of 1836. Most of the cash had been issued by the state banks. Many states had few regulations on the lending and emission policies of their banks. The law permitted the printing of paper money in extravagant amounts. As long as the federal land offices accepted this paper money, speculation was encouraged. The land office sold some 12,564,478 acres in 1835 alone. In the boom year of 1836, a total of 20,074,870 acres passed into private hands.

The Land Law of 1820 thus stimulated land sales, settlement, and astounding growth of farming and farm production. While it was in effect from 1820 to 1841, the act permitted the sale of 87,538,346 acres as compared with 13,647,536 acres sold between 1796 and 1820. Whatever its shortcomings, the law enabled many people to become landowners and aided the distribution of the means of production to those who tilled the land. Such a mass distribution of land was unique in the history of the world, and none had been so effective and equalitarian.

MILITARY LAND BOUNTIES AND THE
MILITARY RESERVE SYSTEM

The Federal Government did not sell all the land it distributed. By long tradition, soldiers had been rewarded with grants of land. Even before the War of 1812, Congress had offered 160 acres to enlisted men who would serve five years in the army or navy. This policy continued throughout the war, but no special bounties were offered to officers during the war. The bounties did not go to men under 18 or over 45, but Congress changed this inequitable provision in 1816. The omission of special benefits to officers proved to be a serious oversight in encouraging recruitment. It also turned out to be politically unacceptable.

Enlisted men had to apply for their land within five years after their service ended, and they had to locate their land within a military reserve. Lottery determined the exact location of the land. Chance thus entered into American land policy for the first time. The lottery did establish precedents for later land laws. Soldiers and sailors received warrants which could be surrendered for the land. The warrants of 1812 and 1816 could not be transferred, so the veteran's warrant was worthless if he had no intention of settling.

The military reserve system with the nontransferrable warrants did not satisfy many or work well either. Its aim was to establish colonies of veterans along the Indian frontier. This projected screen never actually saved settlements from attack, but it may have stimulated encroachment on Indian lands by those who had no military warrants for the reserves. (By 1830, the Indians east of the Mississippi had been intimidated, relocated, or defeated. The Black Hawk War of 1832 represented the last despairing and hopeless resistance of a ruined remnant.)

In subsequent veterans' legislation. Congress adopted even more freehanded disposal methods. Shortly after the Mexican War began, Congress, in February 1847, passed a military bounty act which included all soldiers. The veteran might take treasury scrip for $100 in place of a land warrant, with the scrip to be accepted in payment for any government land. Soldiers, including those who remained in service, could buy land anywhere without any obligation to farm it. Veterans could thus speculate even if they did not want to farm. Neither scrip nor warrants could be transferred, but the land acquired with them could be. The next step was to make the paper transferrable and relieve the veteran of the roundabout method of disposing land he did not want to farm. Congress retroactively made warrants and scrip transferrable in 1852. This act benefitted those who had kept their papers and had not located their land.

In 1850, Congress set up a liberal land grant for soldiers, and the immediate survivors of soldiers. The grants were offered to men of every rank in every service. In 1852, Congress made all previous service land grants assignable so that all entitled to any bounty could receive it. Congress rejected the proposal that the scrip be redeemed by the government for $1.40 an acre if the soldier did not want the land. That proposal would have required an appropriation from the treasury. Soon many agents were offering to buy the scrip of veterans, and ads appeared in newspapers offering to buy or sell scrip. In the North after 1852, farmers purchased a surprising proportion of their land with soldiers' warrants. These could be purchased under the $1.25 minimum bidding price. Soldiers' warrants had the great advantage of not requiring any bidding at auction.

The land acts helped encourage settlement and farming as well as speculation. At the same time, the acts brought no revenue to the treasury. The United States shortly abandoned the concept of making money off the land sales.

Proposals of giving free homesteads had been made early in the history of the Republic. Agitation for such legislation increased in the 1840s, and Congress had moved toward a policy of free land indirectly by way of veterans' benefits. The Act of 1852 established a policy of free land for veterans and cheap land for others. The next step was to enlarge the amount of land thus distributed.

In 1855, Congress offered a bounty of 160 acres of land to any soldier or sailor who had served in any war after 1790. Veterans of Anthony Wayne's campaign in Ohio in 1794, for example, became eligible although many of them were no longer alive to collect their reward. But the heirs of veterans could receive the bounty, and in 1856, Congress extended the bounty to cover veterans of the Revolutionary War or any subsequent war. To qualify, the veteran had to have served two weeks or to have fought in one major engagement. This last provision covered the veterans of battles such as Saratoga or New Orleans where troops had been assembled for one fairly short battle. As before, the heirs (particularly the widows) of the veterans could receive the bounty.

Since Americans had fought wars almost continuously after 1775 and almost everyone had some relative who had served in one of the wars, the act amounted to a general distribution of land and of warrants. As in earlier acts, the recipient could transfer the scrip. Much has been made of the many warrants which ended up with speculators. The Tract Books of the General Land Office, however, show that the bulk of the warrants ended up in the hands of farmers and small-scale farmer-speculators. In short, the act did make landowning easier and less expensive.

FEDERAL LAND DISPOSAL

Counting only the federal acts of 1776 to 1855, the national government disposed of 73,485,399 acres through military bounties. Farmers especially used military bounties in Iowa, Illinois, and Missouri. Over 14 million of Iowa's 36 million acres went to settlers through military scrip and warrants. In all states with public lands, people obtained a sizeable amount of land with bounties. We can scarcely underestimate the significance of the military acts both in terms of disposal and as a precedent for later free land acts. Between 1842 and 1862, the Federal Government sold some 69,198,547 acres of land. The military bounty acts (1847 to 1855) offered another 59,127,730 acres. Some of this land went unclaimed, but the land entered under these and earlier acts was about equal to the amount sold.

The operation of other land laws determined to a large extent the value of the military warrants. Warrant holders tended to seek out the less settled places where they had a better chance of finding high quality land. The farmer had several options other than the purchase and use of military scrip. If the farmer bid for land in an already settled area, his land costs could rise appreciably. If instead he located in sparsely settled country, the bid prices might be fairly low. Still, he might have to bid high for land he had improved. Although claims associations worked well enough, most farmers wanted their titles regularized without forming associations to intimidate other purchasers.

PREEMPTION ACTS

Technically, settlement before purchase had long been illegal. Actually, however, the government had had no effective machinery to enforce the law. Beginning in 1830, Congress, in virtually each session, passed a series of acts which made settlement before purchase legal. In 1841, the Preemption Act allowed farmers to settle legally before they purchased their land. More importantly, the law provided that a settler already on the land could buy up to 160 acres for $1.25 an acre. He did not need to bid at auction, nor did he have to join a claims association to protect his ownership. After more than twenty years of agitation, preemption became a legal reality. Of course, this affected the price of scrip which became more valuable.

The Preemption Act of 1841 also gave each new state 500,000 acres which it could use to finance internal improvements. The states had always been minor landowners because of the land grants for education, but they now received more land. The law also provided that of the land sold, 10 percent of the federal revenue went to the state wherein the land lay. The remainder, minus administrative costs, the Federal Government divided among the states according to population, but Congress ended this distribution in 1842. By then, the worst effects of the depression of 1837 had subsided.

THE GRADUATION ACT

Farmers and speculators in the public land states still felt, in spite of the liberal acts, that the states held land worth less than the minimum price or the value of the military scrip. Senator Thomas Hart Benton from Missouri had for some time pressed for a graduation policy. On August 3, 1854, Congress passed the Graduation Act. Under this act, land which had been on the market for ten years would be reduced to $1 an acre. If the land remained unsold for another five years, the price fell to 75 cents an acre; if unsold for another five years, it would sell for 25 cents an acre; and if still unsold after thirty years, it would sell for 12½ cents an acre. Much of the land so disposed went to land speculators. A large amount of it apparently was used for mining or lumbering rather than for farming as the law intended. The act worked so poorly that it may have speeded the passage of the Homestead Act. In any case, Congress ended graduation in prices on June 2, 1862.

Briefly put, the United States inherited a land policy of mercantilist intent from England in 1783. This policy primarily benefitted the government. By 1862, the Federal Government had worked out a policy of free enterprise in which the state sought to divest itself of its land holdings as rapidly as possible, chiefly for the benefit of farmers. The legislation and practice increasingly took a distinctively agrarian cast. Historians have emphasized the role of land speculators in securing the land laws. With the growth of democratic institutions and the free enterprise system, no one could have put a real brake on speculators without in some way diminishing private initiative. Across the years the laws tended to benefit farmers more than capitalists and extended speculation across a broader spectrum of the population. As time went on, Americans found it easier and cheaper to secure farmland. This democratization of land ownership had a strong effect on the growth of American agricultural efficiency and production.

Many benefits flowed to the American farmer by way of the land laws. Farmers, most significantly, could shift their wealth to capital instead of using it to buy land. The land laws, such as that of 1856, also stimulated the development of transportation routes for the movement of agricultural produce. These internal improvements significantly helped farmers in their drive toward commercial farming. That all acts had some fraud connected with them should not obscure the comparatively honest administration of the laws. Neither should fraud becloud the general assistance the laws seem to have given to farmers in the North, South, and West.

Not that farmers alone benefitted. In the 1840s, reformers began to urge the passage of a homestead act as a method of alleviating the living conditions of the urban worker. George Henry Evans became especially prominent in pressing for the passage of a homestead act, urging it in the *Working Man's Advocate,* the *True Workingman,* and other papers. He and others held that if workers could settle on a farm in the West, then poverty would decline and wages would rise. The theory did not work as intended, largely because even the most primitive pioneering required more capital than the poor could accumulate. The not-so-poor fared better in towns than on the frontier.

The propagandists for free land kept up their pressure, and the first serious homestead legislation came before Congress in 1852. Spokesmen for the workers found themselves joined by reformers interested in the lot of tenant farmers. Both sought passage of a homestead act. But neither western farmers, whose land appreciated rapidly after purchase, nor eastern manufacturers with great political influence wanted the act. The manufacturers believed the act would have the effect the reformers insisted it would have, and that a homestead law would result in a rise in wages. Even so, the House of Representatives passed a homestead act in 1852, but it failed to clear the Senate.

THE HOMESTEAD ACT OF 1862

In the following years, Congress considered other homestead bills but none passed until 1860. President James Buchanan of industrial Pennsylvania vetoed the bill; and enough southerners, representing the planters, opposed the bill so that Congress could not override the veto. There the matter stood until May 20, 1862, when the Republicans had enough votes to push it through and thus fulfill their campaign promise of 1860.

Under the Homestead Act of 1862, any person could file for 160 acres of federal land if he or she met certain conditions:

1. He or she were an American citizen, or had filed his or her intention papers.
2. He or she were 21 years old, or the head of a family, or had served 14 days in the U.S. Army or Navy.
3. He or she had never fought against the United States.

The last provision excluded some Mexicans, Canadians, and Britishers, but mostly excluded some residents of the South. If a farmer met those requirements, he could secure a fee simple title to the land if:

1. He or she had resided on or farmed the claim for five successive years after filing.
2. He or she were then a citizen of the United States.
3. He or she had paid the requisite fees and commissions.

Congress had to alter the law as soon as the Civil War ended and a sizeable group of people could use the act. Both husband and wife could not take out separate homesteads. In 1866, Congress allowed Confederate veterans to take out homesteads. No one could take out a claim if he already owned 160 acres, and farmers could perfect only one claim in a lifetime. Other changes came later as Congress sought to serve farmers and speculators at the same time. The Homestead Act and homesteaders came into conflict with other legislation and other landowners. Congress continually decreased the land available by grants to railroads and to states. These grants had begun well before Congress had seriously considered the Homestead Act, and long before they had passed it.

LAND GRANTS TO RAILROADS

From 1785 onward, the states had received sizeable grants of land, presumably for the support of education. These grants continued and increased in size. Then in 1850, Congress authorized states to make land grants to railroads from the public domain. The Federal Government gave this permission to Illinois, Mississippi, and Alabama. These states had to give the grants in alternate sections of not more than 6 miles on either side of the railroad right-of-way. These three states thus took over an administrative function in the federal public domain. This authority did not continue for long.

In 1851, Congress made the first direct federal land grant to a railroad. The Illinois Central received a subsidy of more than 2.5 million acres of land. In order to secure capital, the railroad effectively and profitably carried on colonization work. The gift of land worked well enough to encourage Congress to expand this method of helping finance a transportation system for farm products. The federal land grant had the advantage of increasing the number of farmers on the land. The next federal land grant went to the Union Pacific and the Central Pacific railroads. On July 1, 1862, these railroads received 10 miles on either side of the right-of-way in alternate sections, but this original grant rose to 20 miles on either side in July 1864. In that year the Northern Pacific received 20 miles on either side of its right-of-way, except in the territories where the railroads received 40 miles on either side of the right-of-way.

All grants depended on the completion of the railroad. If the entrepreneurs failed to build the railroad, the land reverted to the Federal Government. As of 1860, the Federal Government had granted 27,876,772 acres to railroads and had given some 74,395,801 additional acres directly or indirectly by 1865. These gifts reduced the land available to homesteaders. The railroad land grants, in fact, exceeded all the military bounty grants up to that time.

Before passing the Homestead Act, Congress had provided that the land within the railroad grants, which still belonged to the Federal Government, should sell at double the usual price for land. This assured the railroads that they could sell their land first. After passage of the Homestead Act, the government held its sections until the railroad had sold all its land. Even then, for a number of years, a homesteader could secure only 80 acres of the federal land in the checkerboard pattern of the railroad grants. Belts of unobtainable land eventually ran from 60 to 120 miles on either side of the right-of-way. The farmer had his choice of paying the comparatively heavy charges for railroad land or of settling far from the transportation afforded by the railroad. The engrossing of good land by railroads had advanced far even before the Civil War. The railroads had removed from entry about 22 million acres of the best land before Congress passed the Homestead Act.

SUMMARY OF U.S. LAND POLICIES

Most scholarly discussion of land ownership centers on governmental policies and activities. However, most land transactions unquestionably took place privately through sale, inheritance, and lease. Changes by these methods can be charted only in broad outline. The shift from small farms to slightly larger farms took place by way of private transfers; the development of plantations out of a cluster of small pioneer farms also took place privately. Few gigantic landholdings arose from engrossment of public lands. The notable exceptions involved lumber barons and a few cattle barons. Even in these cases, private sales of homesteads served as the vehicle for the accumulation of vast landholdings.

Through private sales and transfer as well as direct purchase from the states and the Federal Government, much of the land of the United States went promptly to individual or corporate land speculators. These speculators usually did not plan to farm the land themselves or to rent it. Land speculators, including the railroads, sometimes profited handsomely, but more often they did not if they failed to get enough control to engage in monopolistic price gouging. The Preemption Law and the informal, but effective, claims associations prevented that.

Unquestionably, in the long run Americans tended to have ever fewer farmers and ever larger farms. Furthermore, from time to time agrarian depressions damaged the economic well-being of farmers. But neither of these developments could be traced to the land policy of the Federal Government. Changes in farming and land ownership would have taken place regardless of the land system. The amount and nature of the changes would have been different under a different land policy, but the land laws could have been much worse—bad enough to have strangled American agricultural development. Public policy, however, never favored the development of fiefdoms. The strict prohibitions against entail and primogeniture prevented the creation of legally inviolate manors and plantations.

From ancient times, land reform has been offered as a cure for the ills of the state. But no land reform in human history has resulted in a division of land which provided justice and equality on a permanent basis. Successful land reform programs have invariably been associated with fundamental changes in capital, farm science, and technology.

The American people and their government followed two general policies implicitly or explicitly. First, the land was to be distributed to farmers, and the number of farmers and farms was to grow as rapidly as possible. This decision was more often phrased in terms of the rights of citizens than in terms of the needs of the nation collectively. As a national policy which led to national strength, the land system did have its defenders.

MANIFEST DESTINY AND LAND ACQUISITION

Second, Americans sought to increase the amount of territory which the nation could use to work out its peculiar system of land distribution. Although farmers could not settle and farm all the available land despite unbelievable migration and immigration, the American people still pursued a policy of adding to the national domain. This land engrossment took place in the 19th century, and most of the land additions occurred between the Revolution and the Civil War. The addition of territory took place under the inspiration of a special spiritual mystique combined with a practical rationalization of the needs of the Republic. The arguments, both spiritual and pragmatic, have usually been lumped under the heading of Manifest Destiny. The term is imperfect and unclear, but long custom requires its use in any discussion of the land policies of the United States.

The national defense also inspired the addition of territory to the United States. The peace commissioners of the United States secured the Old Northwest at the Peace of Paris in 1783. The Americans sought sovereignty over an area from which Indians and British might threaten the United States, but furs, farmland, and other considerations also made the territory desirable. The need for a port to serve farmer marketing needs led to the more or less accidental acquisition of New Orleans and Louisiana in 1803.

The addition of so much land in so short a time and so far in excess of immediate needs raised the prospect in many minds of a single nation, fully settled and bustling with activity, which embraced the entire North American continent. In less than 30 years, the United States had risen from a colony to a great land empire, relatively secure from foreign invasion. The astounding increases in population and the prosperity of American farms could cause anyone to wonder if the country might not shortly emerge as the greatest nation on earth. The notion of being the "last and best hope for all mankind" had been proudly discussed even as the Revolution was under way. The phrase belongs to Tom Paine. Soon after the Louisiana Purchase, a doctrine of national imperialism (though never so called) began to take tentative shape. At first Americans used the idea to explain their successes. In time they acted on the doctrine.

Whatever may be the confused causes of the War of 1812, many Americans certainly did dream of a quick conquest of Canada and Florida. The war failed to produce any such grand success, but the theory and doctrine flourished even more after failure. Manifest Destiny became a popular doctrine for which the Democratic party served as both vehicle and political expression. The slave-ridden South needed land for expansion, and the nonslave North wanted land in compensation and as a balance.

The Mexican War, 1846–1848, produced easy victories, although the campaigns were difficult because of the general confusion and incompetence of the American forces as well as the problems of deserts and mountains and supply. The resultant addition of territory after a brief war (many Indian wars took longer to resolve) enhanced the popular dream that the whole continent would be included in the United States whose boundaries, after all, had been delineated by God and by nature.

The popular belief was that the whole continent belonged to the

United States, and no other people could or would use the rich resources as well as the Americans. Furthermore, the population of the United States was increasing rapidly, and more land would be needed if this rate of increase continued. The issue of slavery diluted the enthusiasm, and eventually abolitionism forced Manifest Destiny into the background. However, destiny seemed to manifest itself in the Mexican War and the annexation of Oregon, whereby the United States increased its total land area by 1,204,896 square miles. American territory comprised a total of 2,997,110 square miles by 1850.

CHAPTER SEVEN

FARM AND INDUSTRIAL MARKETS ⚒ 1783–1860

THE STORY of farm and industrial markets between 1783 and 1860 falls into three periods or stages of development. In the first period (1783 to 1815), farmers mainly exploited the markets developed before the Revolution. These markets became more valuable because of the European wars. Various events started the process of American industrialization and urbanization which strongly influenced farming in the second stage.

The second stage (1815 to 1840) saw westward expansion, transportation improvements, industrial growth, and increasing urbanization. Commercial agriculture expanded. Cotton producers especially prospered, and those who supplied them shared in the prosperity.

In the third stage (1840 to 1860), acquisition of territory and rapid westward movement were accompanied by the agrarian conquest of a large portion of the continent. Simultaneously, American farmers enjoyed greater sales in Europe and in the American cities. Constant improvements in transportation and processing of products facilitated the flow of farm commodities to market. Improvements in technology helped the farmer receive higher returns from his land and labor. A period which promised widespread national farm well-being came to an abrupt end with the outbreak of the Civil War.

CHANGES IN THE FOREIGN MARKETS

The profits in farming helped spur the advance of American farmers across the continent after American independence. The old markets remained and grew, and new ones opened up in the United States and abroad. When their preferential treatment within the British empire ended, Americans lost a few customers, but British needs soon overcame any British desire for revenge.

More importantly, the allies of the Americans furnished markets, at least for a brief period of reconstruction following the war. France opened her West Indian ports to Americans, chiefly to buy foodstuffs. The Dutch always welcomed the Americans. Even the Spanish, uncertain and reluctant allies, allowed trade with Cuba. Four European and African nations signed commercial treaties with the United States. These treaties were chiefly for farm products as Americans had little else to export. Even the British soon allowed direct American trade with their islands in the West Indies with no duties on certain American farm products. The British and others actually had no other options if they wanted to maintain their sugar islands in full production.

The islanders had long known that food from Europe cost far too much, and that they had no ready source of food or lumber except America. Although the West Indian planters were the best foreign customers for food, Americans traded everywhere. Southern Europeans bought a great deal of grain, and Europeans generally bought sizeable quantities of tobacco.

Substantial changes in the European market took place after 1793. The huge armies demanded by the Napoleonic Wars were the largest in Europe since Caesar Augustus, and they drained Europe of farm workers. In most places farm production either declined or failed to rise to meet military needs. The destruction of war took its toll long before the Russians burned their fields in the face of the Napoleonic armies. The needs of the armies diminished supplies for the general population, as always happens in war. The British, though not on the Continent, suffered the injury of the loss of trade with the Baltic countries; Napoleon closed the northern ports and denied the British a key source of grain and lumber.

The British, because of their better control of the seas, rapidly became the prime market for the foods of America. From 1784 to 1793, the price of flour at Philadelphia averaged only $5.41 a barrel, then rose sharply to an average of $9.12 a barrel between 1793 and 1807, and fell to an average of $5.46 from 1807 to 1816. At least through half the Napoleonic Wars (1803–1815), the American grain growers experienced a delightful prosperity. French and British blockades and other economic actions of the belligerents finally ended it.

In 1793, as the great wars began, Eli Whitney invented the cotton gin. Soon American cotton was figuring importantly in European trade, and it amounted to 30 percent of the U.S. farm exports by 1805. This lucrative trade and the resulting farm prosperity was damaged by the British Orders in Council of 1804 and 1806 which placed most of Europe under blockade. Then came the Napoleonic decrees of Berlin and Milan which closed the Continent to trade with England or to any ships which obeyed the British Orders in Council.

Shippers and merchants suffered most from the acts. The French and British seized about 1,600 U.S. ships during this period of economic warfare. The belligerants confiscated about $60 million worth of cargo, most of it agricultural produce.

In reply, the Americans instituted an embargo from 1807 to 1809. The law cut all trade drastically. In one year, exports fell from $108,300,000 to $22,400,000, or nearly an 80 percent decline. Farmers suffered most in the loss of the European market. Many farmers went into bankruptcy because they had borrowed heavily to expand production during the period of high prices. The cotton planters and farmers of the South had borrowed chiefly to buy land and slaves. As the European market disappeared, the loans could not be repaid. Agricultural items, except dairy products, had to move in international commerce to give farmers any significant returns. The American metropolis could not absorb the total American farm production at prices profitable to farmers as early as 1800. This included even relatively nonperishable and storable products, such as tobacco and cotton. The embargo ended in 1809 and trade returned on British terms. Another commercial–agrarian boom was under way by 1811.

The boom resulted in part from successful American defiance of the British Orders in Council and the running of blockades. However, the British determined to stop these food shipments to France and the rest of the Continent, to end this aid to their enemy, and to secure needed commodities for themselves. The success of the British—who seized ships and cargoes with impunity—once again threatened American commercial and farm prosperity. Furthermore, the British seizure of American seamen under the pretense they were British citizens did nothing to improve relations.

WAR OF 1812

In 1812, the United States went to war with Great Britain. The varied causes of the war included American determination to crush the Indians, conquer Canada and Florida, and end impressment of seamen. But an agrarian desire to shake off British trade restrictions and to open up more markets for American farm produce undoubtedly influenced many to support the war. In starting the war the Americans, in effect, sided with France against Great Britain. But no formal agreements were made, and the Americans generally disliked the association.

America realized none of its immediate aims of the war. Canada did not fall, nor did Florida, and the British navy effectively cut off American trade, blockaded ports, and captured over 1,400 merchant and fishing ships. When the Americans went to war, Napoleon was beginning his disastrous invasion of Russia. The absence of Napoleon left the British comparatively free to deal with America. Foreign trade fell to the lowest levels in American history, and once again American farmers suffered for want of markets. Moreover, westward migration virtually stopped because of Indian uprisings on the side of the British. Some exposed settled areas were raided by Indians or the British, including the capitol at Washington.

But the British could win no decisive battles, and the tide began to shift slowly but perceptibly to the American cause. Meanwhile, Napoleon returned from Russia, reorganized, and struck again at his adversaries, particularly the British. The British needed to end a perplexing and useless war in America so they might turn their attention to Napoleon. The diversion caused by the French helped the Americans. Although the war ended in December 1814, the news came too late to save a British army at New Orleans from decimation.

The end of the War of 1812 coincided with the end of Napoleon for all practical purposes. However, the return of peace in America and in Europe meant the drying up of the formerly profitable trade in farm commodities except for a small but growing market for cotton. By 1815, cotton amounted to 45 percent of the value of all exported farm products and rose to 54 percent of that total over the next five years.

The war had forced Americans to increase manufacturing because of their inability to import supplies. During the war, capital, which might otherwise have gone into shipbuilding, commerce, or land speculation, had to be employed elsewhere in manufacture. Not surprisingly, the first real industrialists in America were merchants formerly engaged in foreign trade. Land speculation during a general Indian war did not offer good returns. Commerce they found impossible. The result was the slow but certain growth of industry and with it the greater growth of cities. Already by 1815 the American metropolis gave promise of consuming the products of American farms, except for tobacco and cotton which still remained dependent on the European market.

During the Napoleonic Wars the British had passed corn laws which put a tariff on foreign grain (called corn in England) whenever the price in England fell below a certain level. The laws reflected the political power and economic policy of an aristocracy which still had an agrarian base. At the time little opposition was voiced because of the abnormally high price of grain during the war. Foreign grain came in virtually tariff free. After the war and the fall in grain prices, the corn laws kept foreign grain out and the price of bread high in Britain. The act of 1815 had more impact on British domestic politics than it had on the U.S. wheat growers. Still, wheat exports to Europe did fall in the years from 1816 to 1846. But the elimination of this wheat market during these years may have stimulated southerners to grow cotton rather than wheat. Even the devastating Civil War did not halt cotton growing in the South.

INCREASING AGRICULTRAL EXPORTS

On the whole, American agricultural exports rose throughout the 19th century. Industrialization and urbanization in Europe increased demand for American agricultural products. Americans chiefly exported industrial materials and amelioratives rather than food. Cotton became the main export, and it had already passed tobacco in importance by 1803. Between the Congress of Vienna (1815) and the repeal of the corn laws (1846), food exports to Europe did not rise much. The British provided the greatest European market for food in the 19th century. Other people bought American food, however, mainly the West Indians. Every year up to 1830 agricultural exports to Europe rose, but they shot up spectacularly in 1835 and again in 1847. In the latter year, cotton fell below foods in value of exports for the first time since around 1810. Otherwise the value of cotton exceeded that of all other agricultural exports combined.

The development of internal transportation in the United States helped speed internal as well as international commerce. The completion of the Erie Canal in 1825 allowed a food surplus from the North and Old Northwest to be sent to the urban East and the overseas markets. The development of steam transport on the Mississippi ranked as the chief innovation. It began in 1812 and became significant after 1816. Now food could be moved to cotton plantations and the cotton shipped out much faster.

The new, fast clipper ships enabled the United States to increase agricultural exports in the 1840s and 1850s. After 1847, the development of railroads, ocean shipping, and the expanding European market allowed American farmers to sell their surpluses to the world.

After the Mexican War, food exports held up well and cotton exports shot up. Cotton made up 60 percent of all agricultural exports in 1849 and rose irregularly to 75 percent of all farm exports by 1860. Animal foods and products fluctuated as a percentage of the total of farm exports, but advanced steadily in absolute terms. Grain, particularly wheat and flour, made up the bulk of the food exports. Grain regularly amounted to two or three times the value of animal products. As late as 1860, manufactured goods averaged only 12 percent of the total exports. Farming was the basic industry of the Republic, and cotton its chief commercial commodity. By the late 1850s, cotton exports exceeded the total value of all other exports, agricultural and manufactured, combined.

As trade with Europe increased, trade with the West Indies and other places decreased. The repeal of the British corn laws in 1846, crop failures in Ireland from 1845 on, and English crop failures in 1850 all stimulated trade with Europe. The Crimean War (1854–1856) and the gold rushes in America and Australia also produced new markets for American goods which were overwhelmingly agricultural.

In spite of steady industrial advance, the United States stood only a weak fourth on the list of industrial nations, judged by the value of its manufactured products. In addition, most American manufactured goods went into domestic consumption. The growing manufacturing industries of the United States were financed largely through the profits from agricultural exports. These profits paid for the capital and the talent needed for industrial expansion. By 1860, Europeans bought some 75 percent of all American exports, mainly cotton. Cotton and tobacco provided the surplus capital needed to develop the American transportation systems. Domestic capital, although small in amount, was thus released for the development of manufacturing.

NEW DEVELOPMENTS IN MANUFACTURING

In several industries, notably textiles, the Americans adopted English manufacturing methods and improved on them. Technically, but not in volume or in value, American manufacturers soon surpassed their European competitors. The shortage of labor largely explained American efforts to use machines whenever possible. However, the machines created more jobs than were eliminated. Even before 1790, spinning machines had appeared in Philadelphia and several cities in New England. But the big breakthrough came when Samuel Slater introduced Arkwright's machinery into Rhode Island in 1791. The embargo of 1807–1809 and the War of 1812 had stimulated the building of factories to supply American needs. Not only cotton but woolen mills increased markedly, especially after 1792 and between 1812 and 1815.

In the 18th century, standardization of parts in manufactured items began in the United States. As machine tools became more exact, the practice spread to other industries. No later than 1817, Jethro Wood patented and began to make iron plows with interchangeable parts, although the principle had been worked out as early as 1813. The jig, or guide for repeated cuttings, came to be used extensively.

It facilitated the mass production of clocks and various farm machines and implements. The limited innovations in farm machinery made in the 18th century were more widely adopted and rapidly improved. Industrial advances in technology quickly influenced American farming by making possible the production of effective farm tools and implements more cheaply than ever before.

The new small industries also afforded more and better markets for the American farmer. The industrialization of the textile industry notably increased the market for cotton, but many other industries benefitted from such advances. The use of the sewing machine in shoemaking, for example, had an impact on the hide and skin business. More people could afford more leather goods, and the demand for hides and leather rose more than a simple increase in population would have required.

The rate and nature of industrial changes varied, but an overview of American manufacturing is possible. From 1783 to the Civil War, the chief industries in order of importance were flour and meal milling; cotton textile manufacturing; wood and wood products; iron smelting and manufacture; shoe manufacturing; clothing; leather tanning and hide preparation; woolen textiles; sugar refining; and food processing.

Except for iron manufacturing, every major industry directly used some agricultural product. Every industry experienced technological developments, either in new devices or larger scales of operation. Processing changes always influenced farmers, although their impact sometimes eludes easy assessment. On the whole, the farmers benefitted from them.

TECHNOLOGICAL ADVANCES IN THE COTTON INDUSTRY

In the South, cotton occupied the same position of importance as bread grains held in the North. In the late 18th and early 19th centuries, technological changes aided the cotton industry. Toll gins were widespread in the South by 1796, used primarily by small farmers. The big planters preferred to use their own gins. Before the Civil War, most ginning took place on the plantations. For the most part, the application of power remained primitive, but users of gins gradually made other improvements. Better methods for removing the trash and dirt yielded a premium quality of cotton. By the 1840s, each gin needed about three people to process from $1\frac{1}{2}$ to 6 bales a day. By the 1860s, many planters owned two gins which could gin from 300 to 500 bales a day.

In the early days, from 1783 to about 1800, planters sacked their cotton. They pressed it in by foot or packed it down with a hand tamp or plunger, taking care not to injure the fibers. This expensive method of packing and pressing survived into the 1840s in out-of-the-way places. It took one man a day to fill a sack of cotton this way, and the resulting package often weighed as little as 250 pounds.

With a screw press, two men and a team of mules could pack around 14 sacks a day. The lever press had appeared first; but the screw press, with a sweep attached to mules or horses, generally superseded the lever press by about 1801. The sacks were usually made of hemp, and the bales were tied with hemp rope.

Factors and merchants soon entered the pressing business, however, because they discovered that most planters did not press the cotton tightly enough. Ocean transport charges were based on volume rather than on weight, and most cotton went by sea to Europe or to northern mills. Therefore, reducing the volume of the cotton before shipping was essential in cutting transportation costs. Every port city had presses for repressing cotton prior to shipment. This system of second pressing continued right up to the Civil War. The merchants, brokers, factors, and others also had an interest in uniformity of bale weight in order to make bookkeeping easier. Merchants in control of the terminal presses standardized the bale weight at 500 pounds.

After the cotton had been ginned and pressed, it had to be stored until shipped. So the factors and brokers built large, fireproof warehouses. For example, by 1851, Mobile, Alabama, had 42 fireproof brick warehouses with a total capacity of 310,000 bales. Other ports had similar facilities. Cotton could be moved more easily with hooks and block and tackle than most other goods. Even so, the 500-pound bales had to be manhandled along the way, usually by slaves. The bale weighed too much to be lifted by one man, but two men could roll a bale much as they could roll a barrel.

Cotton culture spread into the interior of the South, much as wheat spread in the North. The size of the cotton crop consistently rose. The increasing profits arose primarily from the widening of the market as the result of improvements in textile manufacture and transport systems. However, eastern cotton producers soon faced the same difficulties that the eastern grain farmers faced—declining profits in competition with western producers who had easy access to European markets. Declining soil fertility also hurt the easterners. Instead of changing crops, planters of the Southeast responded with attempts at increasing the size of plantations. They survived, but cotton profits were less than for other crops by 1850. Still, technological improvements in ginning, pressing, and transport did measurably increase the profits of planters in Alabama, Mississippi, Arkansas, and points west.

TEXTILE INDUSTRIES

The big increase of cotton textile mills began after 1812. The United States had 795 mills worth $45 million by 1830; the number of mills increased to 1,240 by 1840. Most cotton textiles used in 1840 were of domestic manufacture. Even so, the textile mills used only a small percentage of the cotton crop in the 1840s. On the other hand, the woolen industry was hurt during the 1820s by foreign competition, and importations rose in spite of tariff protection. The tariff and industrial advances helped the woolen industry in the 1830s, and the United States had 1,420 woolen mills by 1840.

Both the woolen and cotton industries suffered in the depression of 1837–1842. The woolen industry, however, recovered rapidly in the 1840s. Farmers of New England and Ohio especially responded with a heavy wool production which mostly went into American textiles.

Ready-made clothes were first produced in volume in the 1830s. They provided not only a new industry for the cities, but another market for the farmers. The invention and improvement of sewing machines reduced the costs of and opened wider local markets for ready-made clothes. Advances in textile production and technology thus further expanded markets for fiber producers. In contrast, food producers faced, sooner or later, a glutted market. This problem never became serious because population continued to increase rapidly.

HANDLING OF OTHER FARM PRODUCTS

Farmers raised more than cotton, wheat, corn, or livestock. A host of other farm products had to be sold. Handling perishable commodities required great care and work; this raised their costs. Fresh produce could be consumed in season and, like strawberries from Florida, might be advanced or retarded in season by the distance grown from the market. But the greater distance added to the cost. Perishability made both processing and transportation cost more. For consumers, the greatest difficulty was getting foods out of season, regardless of cost.

CHANGES IN FOOD PROCESSING

American flour millers industrialized early. In 1841, Joseph Dart of Buffalo, New York, began to use grain elevators to unload canal and lake boats. His elevator consisted of buckets on a continuous chain. The device added tremendously to the handling capacity of millers and also helped centralize the industry. Substantial advances in milling technology came later, but uniformity of product was already appearing. By the Civil War the millers of Buffalo used elevators to handle volumes of grain which would have been impossible using available workers and the old bag system. Centralization of milling also reduced the transport needs of the industry and cut the handling costs for wheat and flour. The village miller, who had formerly controlled marketing and prices within his territory, gave way to the wider world market. The change brought better prices to the farmers.

Other advances took place in food processing. Canning was introduced before 1840 on a small scale, and a more centralized, better capitalized food processing and distribution had begun. In milling, canning, and other related industries emerging before 1865, development hinged on the availability of good, cheap iron and steel.

NEW TECHNOLOGY IN THE IRON AND STEEL INDUSTRY

In almost every phase of the agricultural enterprise, technological changes in the iron and steel industry had immediate and important results. The anthracite blast furnace, introduced in 1833, greatly cut costs and produced new efficiencies in making iron. The new furnace allowed the industry to operate independently of charcoal. Few farmers were affected by the loss of the charcoal market. The new blast furnaces came into wide use during the 1840s; and with the end of the dependence on charcoal, they were usually built in urban areas. By about 1855, steelmakers had shifted to anthracite smelting, although a large minority of producers still used charcoal. Coke as a fuel did not become more important than coal until after 1865. The steel town had begun to appear by the late 1840s.

The demands of industrialization called for many large castings. The water turbine, which replaced the water wheel, required cast iron instead of wood. Factory machinery formerly was made of wood, which was replaced by cast iron. Steam engines, often running on coal, powered the new machinery; and industry, freed of dependence on water power, clustered in or near larger urban concentrations. The heavy casting industries in the 1850s were clustered at such places as Wheeling, Pittsburgh, Cincinnati, New York, and Providence because their river shipbuilding industries provided the surest and steadiest market for the heavy casting of steam engines.

In 1844, American manufacturers, using rolling mills, began to make iron rails for railroads and American dependence on foreign iron and steel began to decline. During this period of advancing iron

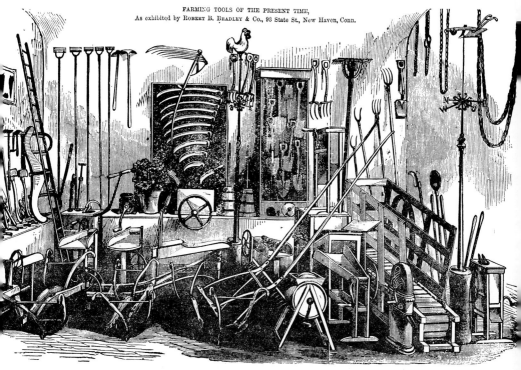

FARMING TOOLS OF THE PRESENT TIME,
As exhibited by ROBERT B. BRADLEY & Co., 93 State St., New Haven, Conn.

Farming tools, *ca.* 1861. (From *Eighty Years of Progress of the United States,* Vol. 1, New York, 1861)

technology and cheaper products, steel and iron came to be used more extensively and successfully in tillage implements, notably the plows of John Deere. Gradually they replaced wood in most machinery.

Concurrently, machines came into use for making a variety of small articles—from nails and spikes to files and chains. Although many of the machines actually had been invented earlier, they did not become economically significant until after 1840. Centralization and mass production helped destroy the minor monopolies of countless small blacksmiths and the hardware store came into its own. The continued development of machine tools, designed for mass production and interchangeability of parts, put Americans ahead of Europeans technologically by 1850. Business organization and factory design and layout played as important a role in advancing technology as did inventions and engineering. While these technological developments were taking place, the seaboard towns continued to serve farmers both as urban markets and as distribution points. A large amount of the agricultural surplus of American farms went abroad. Imports of manufactured goods still ranked high in economic activity between 1815 and 1840 for Americans were by no means self-sufficient.

The volume of trade grew between 1815 and 1860. No doubt

much of the expansion in the 1830s and 1840s stemmed from favorable tariff legislation. Undoubtedly, farmers generally bore the brunt of the tariff which resulted from Henry Clay's American System. Clay proposed that American industry should be developed to provide farmers with a rich domestic market. The system worked, but Congress never actually endorsed it as public policy. Although farmers had to pay more duty than they might have without a tariff, they did receive an ever larger domestic market. The foreign market was still important for commercial farmers and existed despite the tariffs. However, it meant a dependence on the general economic and political fortunes of foreigners, which gradually declined as American industry and cities expanded.

INCREASED URBANIZATION

More urbanization accompanied the growth of both commerce and manufacturing. After the War of 1812, urban growth increased impressively, mostly by the drift of country people to the towns. Immigration probably accounted for very little of the increase until much later. No one kept data on immigration until after 1819. However, a generous estimate places the number of immigrants well below the number of people moving to the cities. Between 1790 and 1819, some 234,000 people came to the United States, while urban areas grew by about 419,000. Total population rose from 3,728,000 in 1790 to 8,945,000 in 1820. People in cities increased from 5 percent of the total in 1790 to 7 percent in 1820. Urban population rose from 693,000 in 1820 to 3,544,000 by 1850. That is: in 75 years the urban population alone exceeded the total population of 1775. Rural and farm population rose from 8,945,000 in 1820 to 19,648,000 in 1859. The percentage of urban population increased from 7 percent of the total in 1820 to 15 percent by 1850. In each succeeding decade the percentage of urban people increased over the increase of the preceding ten-year period. By 1860, the number of urbanites rose to 6,217,000; from 15 percent of the population in 1850 to nearly 20 percent in 1860.

MASS PROCESSING OF FARM PRODUCTS

Mass processing of farm products developed in response to the needs of these people. In some cases, the improvements concerned management concept more than technology. Meat packing is a case in point. Cincinnati butchers and packers had started processing pork as early as 1818. At first farmers brought their hogs in already slaughtered, but later the packers began paying premiums for live hogs. The development of the assembly line slaughter of hogs in large volume allowed the better use of the total carcass. Cincinnati packers had fully developed volume slaughter and full use of the carcass by 1848 when they packed nearly 500,000 hogs.

Chicago began to grow after 1848 with the opening of the Illinois and Michigan Canal. By 1860, processors in Chicago had begun to challenge those in Cincinnati. A shift of farming population, the advent of the railroad, and some urban market changes soon made Chicago the major meat processing center in the country.

Similar changes and centralization took place in the processing of wheat and other grains. Extensive use of water power, greater urban markets, and the availability of water transportation, combined with nearly fully automated mills, helped make Baltimore, Maryland, Buffalo and Oswego, New York, and St. Louis, Missouri, into flour milling centers. Between 1850 and 1860, Baltimore annually handled some 500,000 barrels of flour. Rochester, New York, grew as a milling center after the opening of the Erie Canal in 1825 and drew on western New York, particularly the Genesee valley, for its wheat supply. In the 1850s, the mills of Rochester produced more than 500,000 barrels of flour a year. The flour and grist mills of America in 1850 produced $136 million worth of flour and ground grain. This made processed grain the most important industry in the country, judged by the value of the product. Flour and kindred products surpassed in value even cotton textiles and dwarfed the value of other manufactured or processed items.

Another agricultural processing industry of increasing importance was shoemaking. The industry, although centrally located, operated under the putting-out system, managed by merchant craftsmen. Some of these industrialists had fairly large but not truly centralized operations. Most of the merchant craftsmen failed or declined during the depression of 1837–1842, and a new system arose to replace them. By the 1840s central factories with systems of standardization took hold. In the fifties, shoes made to fit either right or left feet, not interchangeably, came to predominate.

In the value of the final product, the shoe industry ranked fourth in the nation between 1850 and 1860. Hides came from everywhere, and the most remote livestock raiser could realize something on his hides even if his meat or dairy products were not very profitable.

Living standards rose markedly because of processing efficiencies, and this in turn helped create wider and more certain markets for the farmers who produced the raw materials. The changing industries concentrated in the cities, and so provided wealth for additional workers to consume the various products of the farm. Before 1820, urbanites served less as direct consumers for farmers than as redistributers for the overseas trade. But the city-farm relationship changed slowly but importantly between 1820 and 1865. By the end of the period American cities had achieved a position of market dominance for most farm commodities. However, exports still gave commercial farmers their margin of success.

WESTWARD MOVEMENT OF FARMERS

As farmers moved westward, those remaining in the East complained of hard times brought on by competition with western rivals. The complaints had little substance. Although not often remarked on, western farmers actually forced changes in efficiencies and in commodities on eastern farmers. Westerners used some of the industrial goods of the East, such as iron tools and implements, and with this capital input increased their own effectiveness and output. The rapid development of western towns helped make commercial agriculture more common in the West. The shift of population, with its urbanization, explained many of the developments of American farming everywhere.

At the end of the Revolution, possibly 94 percent of Americans lived east of the Appalachians. A few western settlements in North Carolina, the area that became Tennessee, and Kentucky, plus a few farmers in western Pennsylvania and Maryland accounted for most of the population west of the mountains. The land and governmental policies of the United States encouraged great numbers of farmers to move westward, especially after 1787. In quick succession new states entered the Union: Kentucky in 1792, Tennessee in 1796, Ohio in 1802, Louisiana in 1812, Indiana in 1816, Mississippi in 1817, Illinois in 1818, and Alabama in 1819. Most people in these states farmed. For the most part, those of the northwest raised food and those of the southwest raised cotton.

East of the Mississippi, most of the land was forested. Still, pioneers discovered oak openings and prairies that occurred in both the north and south. West of the Mississippi, however, the scattered prairies gradually merged into one large prairie that reached the rising elevation of the Great Plains. The northern areas of mixed woodland and prairies, just north of the Ohio River, had been glaciated and the limestone mechanically ground into the soil. This area of fantastic fertility needed more rain but still formed one of the garden spots of the world. To the south, the limestone soils of Kentucky and Tennessee and the flood plains of the rivers proved astonishingly fertile. None of these quite matched the magnificent Black Belt of

Alabama, a rich limestone soil of great fertility. The crescent-shaped belt conformed to the terminal arc of Appalachians. The Delta of the Mississippi also stood in a class by itself. Through eons the unparalleled topsoil of the glaciated area to the north had eroded and washed downstream and deposited to a great depth in Tennessee, Missouri, Arkansas, Mississippi, and Louisiana. Farmers had settled all of these areas between 1815 and 1848.

As the grain and meat animals of the North and the cotton, sugar, and rice of the South began to enter the markets in quantity, farmers of the Old Northwest, Middle Atlantic, and the Southwest all felt the pinch of competition. They slowly shifted crops in most of the East, but generally inadequately to meet the requirements of a changing market. The northern and middle states developed cattle feeding, dairying, and vegetable and poultry raising. The Old South, in addition to cotton, found new revenue in selling slaves to the expanding West. Not all migration of farmers brought immediate market consequences. Only those areas with easy water transportation proved commercially valuable until the railroads or canals opened the rest of the territory to the egress of agricultural products and the ingress of manufactured goods. Until then, farmers drove their mobile products of hogs, sheep, cattle, and slaves to market.

Starting in 1849, gold rushes brought miners west to California and later to other places. The migrating fortune hunters formed an important market in places where none had been before. Cattlemen in various places found unexpected customers. Many miners who failed turned to farming, and some made their fortunes by selling food to their mining colleagues. The mining rushes also caused such great, if sometimes temporary, increases in population that new states were formed. Miners often clustered in urban places; some of which later became ghost towns. However, San Francisco and Sacramento stand as obvious examples of permanent settlements. These western urban centers sometimes provided markets for eastern farmers. Entrepreneurs profitably shipped ice from the east coast to the west! The gold discoveries inflated the money supply by simply increasing the gold in circulation. So, westward expansion not only stimulated the development of farming, but provided markets and money.

CHAPTER EIGHT

TRANSPORTATION: MOVING THE PRODUCTS
FROM FARM TO MARKET ✗ 1783–1860

N
OWHERE did the new technology have more significance for the farmer than in transportation—the key to marketing. Changes which speeded the transport of products, made their movement possible, and reduced the cost of transport or handling all had an impact on farming, although not always a favorable one. A simple line of transportation, such as the Erie Canal, not only opened a market for the grain of the Old Northwest, but also progressively ruined the grain farmers of New England, New York, and Pennsylvania. Western grain, shipped quickly and cheaply by water, could now reach European markets. There the influx lowered the prices received by European peasants.

Importation of cheap American grain played some role in stimulating emigration of European peasants to their own cities. This clustering of unskilled and inexpensive labor stimulated industry in the cities of western Europe particularly. Increasing European industrialism created more markets for American agricultural produce. River steamboats, turnpikes, canals, and especially railroads could now all be used to speed transport at lowered costs. This transport improvement led American farmers to produce heavily for the urban markets at home and abroad.

In the resulting competitive squeeze some farmers were forced into different forms of commercial activity. Sometimes they shifted commodities; some changed from wheat to dairying. Cities grew by immigration from rural areas, not from natural population increase. In New England and the South Atlantic regions, farmers sometimes abandoned their farms and moved farther west. Nearly 40 percent of those born in South Carolina and Vermont had left their native states by 1860. Records in the older states showed around 30 percent of those born in any state had gone westward by 1860. In all changes, the improvement of transportation and the resulting dislocations played a significant part.

Throughout the years 1783 to 1860, the European market (sometimes called the world market) dominated the patterns of American agriculture. To a great extent the world market determined what American farmers grew, as in the case of cotton. Reaching this market was vital to American farmers.

DEVELOPMENT OF OCEAN TRANSPORT

Many substantial changes took place in the development of ocean transport between the Revolution and the Civil War. Immediately after the Revolution, the British carried most American goods, but the French also carried a great deal. However, the Napoleonic Wars, beginning about 1793, not only gave American farmers a new market, but eliminated the French carrying business and seriously reduced that of the British. Americans quickly moved to fill the gaps in the carrying business. By 1800, some 92 percent of all commerce between America and Europe moved in American ships. Furthermore, the Americans engaged in the carrying trade for both the English and the French. This in turn stimulated American shipbuilding and advances in American marine architecture.

Most American tonnage (a shipping ton is 100 cubic feet of space) carried the coastal trade. But the number of American ships in foreign trade rose from 124,000 tons in 1789 to 981,000 tons in 1810. By 1860, out of a total merchant tonnage of 5,354,000, foreign trade engaged 2,379,000. However, the American shipping had declined relative to European tonnage.

The War of 1812 set back American shipping, but the introduction of the clipper ship allowed it to regain its prewar position by 1845. American carriers increased consistently thereafter because of inexpensive materials, notably lumber, and the considerable skills of the shipbuilders. The increasing use of iron and steam, however, left wooden vessels at a disadvantage.

Although the clippers were faster, cheaper, and safer than the earliest steam vessels, steamships eventually replaced sailing ships. The development of the screw propeller placed the clipper at a decided disadvantage. By the 1850s, the propeller ships had clearly shown their superiority, and the American clipper—and the American merchant marine—began a slow decline. British subsidy of their merchant marine helped them in the competitive struggle. By 1860, only 66 percent of American foreign commerce moved in American ships, compared to 92 percent in 1826.

The first ocean steamship, the American built *Savannah,* had successfully crossed the Atlantic in 1819. But the British, rather than the Americans, pushed ahead with the use of steamships. In 1838, the wooden paddle-wheelers, the *Sirius* and the *Great Western,* ran under steam from Britain to America in a little over two weeks. Large iron ships came into their own when the first propeller-driven ship, the *Great Britain,* crossed the Atlantic in 1843.

Throughout the period, the trend had been toward regularity of sailing schedules and increasingly safe passages. Propeller-driven ships could keep better schedules in rough weather. But, by wind or by steam, the cost and the time of passage had been gradually reduced. Thus farmers ever more distant from the European market could reach it profitably. The revolution in ocean transport put the wheat farmers of Illinois incomparably closer to the docks of Liverpool. This aided American farmers, but hurt European farmers. The Americans had the advantages also of more land and better technology which fast and inexpensive transportation allowed them to exploit. For farmers in the interior, the trouble and expense of getting their products to a port exceeded their difficulties in getting them to Europe. Improvements in internal transportation actually had more influence on American farmers than improvements in ocean shipping.

DEVELOPMENTS IN ROAD BUILDING

The first major development in road construction and use was the creation of privately owned turnpikes. The first of any length ran between Lancaster and Philadelphia, Pennsylvania. Completed in 1794, this pike turned in such high profits that a turnpike mania hit the country on a scale not equalled until the 20th century! New York had chartered 137 turnpike companies by 1811, and the New England states had chartered another 200 companies. By the 1820s, entrepreneurs had apparently overbuilt turnpikes. The roads exceeded the needs of the users, so the owners tried to sell them to the states. The states not only bought the roads, but continued to build them until the depression of 1837–1842 ruined state finances.

Technologically, road building advanced little in most of the United States during this turnpike era. The federal specifications for the National Road ordered that stumps no higher than 18 inches be left in the roadway. The provision suggests the level of performance obtained. Macadamized roads were built in the eastern states in the 19th century. However, in many places plank roads became more popular in the 1840s and 1850s. Difficult to maintain, these were used mostly to connect towns with railroad centers. Corduroy roads, made of logs placed at right angles to the roadbed, were common where wood was plentiful. In treeless areas, most roads were dirt throughout the 19th century.

State and private efforts were soon supplemented by federal road building. The main incentive came from the West because inadequate transportation kept farmers from moving their produce at a profit. In 1806, Congress authorized the building of a road westward from Cumberland, Maryland, with a terminal in Illinois. Construction on the National Road began in 1811, and the contractors completed the road to Wheeling on the Ohio River by 1818. They would have built it faster except for the disruption caused by the War of 1812. The Federal Government surrendered control of the National Road to the states during the 1830s, and the states sometimes made their portions into toll roads. The U.S. Government helped in building other roads, but it built only the National Road entirely with federal funds.

In spite of sometimes poor construction, the National Road sharply cut wagon transit time for the trip from Baltimore to Wheeling. Time in transit fell from eight days to three days with consequent savings in freight charges. Initially the road had more impact in helping move capital goods to the west then in moving farm commodities to the east. Still, drovers trailed large numbers of livestock, cattle and hogs, from west to east on the road between 1818 and 1865. An important service, which was provided for the drovers, was the building and maintenance of feeding stations, pens, and taverns at regular intervals. These accommodations occurred about 15 miles apart, or an ordinary day's walk with animals.

The surge of western farm products, particularly animals, had almost immediate repercussions on eastern farmers, food processors, and merchants. Trailed cattle and hogs had tough meat. Farmers near eastern cities soon found they could buy western stock to fatten and then sell it at a profit. Thus the feeder industry began in Pennsylvania, New York, Virginia, and New England.

CANAL BUILDING AND WATER TRANSPORT OF PRODUCTS

Between 1783 and 1800, many small canals had been built—mostly to get around waterfalls. More ambitious projects had been proposed. The one most commonly discussed was a canal to run between Albany and Buffalo, linking Lake Erie with the Hudson River. In 1817, the legislature of New York, unable to obtain federal aid, allocated money for building the 363-mile Erie Canal. Work began in 1817 and the canal was completed in 1825. In ten years the tolls had completely repaid the cost of construction, and thereafter the tolls simply produced revenue for the state. The canal had an immediate agricultural impact since it linked the regions of farm surplus in the West with the metropolis of New York, and by extension with the world market. The canal also became a heavily traveled route for farmers moving from New England and New York to the fertile lands of the Old Northwest.

Both time in transit and cost of transportation declined markedly for the trip between Albany and Buffalo. The time fell from twenty to six days, and the cost fell from $100 a ton to $10 a ton. The benefits worked both ways: western farmers had a new market for their produce, and they saved on the cost of manufactured items they used. Both domestic and foreign manufacturers moved to the upper Mississippi valley.

New York also built canals feeding into the Erie Canal. The tonnage figures show the pattern of tremendous growth. In 1837, the Erie Canal alone handled 667,000 tons, and all New York canals handled a total of 1,171,000 tons. By 1850, this volume reached 1,635,000 for the Erie Canal and 3,076,617 for all New York canals, and by 1860, it rose to 2,254,000 and 4,650,000 tons respectively. Part of the tonnage on the Erie Canal came from feeder lines built in other states which joined by way of Lake Erie. Ohio built an elaborate system of canals. In 1828, Kentucky and the Federal Government built the Portland Canal around the falls of the Ohio River at Louisville. Canada in 1833 built the Welland Canal around Nigara Falls.

In general the canals shifted the movement of grain and other commodities away from the Mississippi River and New Orleans toward New York City. This proved to be advantageous for the farmers because the Ohio-Mississippi route had several disadvantages. Until 1828, boatmen had trouble negotiating the falls of the Ohio. The seasonal dumping of farm commodities on the market at New Orleans resulted in lower prices for farmers than more even marketing would have produced. Furthermore, the humid climate of New Orleans hastened spoilage of grain. Because East Coast cities provided the chief markets, the grain still had to be transshipped from New Orleans. By the time all these transportation costs were paid, farmers and merchants had lost an appreciable amount of income to the supposedly cheap water transport.

In 1818, when flour sold at $8 a barrel in New York, it sold at Cincinnati for $3.50. Roughly $4.50 of the $8 went into transportation costs on rivers and coasts. By canal the savings amounted to nearly $3 a barrel. No wonder that the canals stimulated westward migration and encouraged western farmers to secure more land and grow more. Although production, particularly of grains, was increasing, the market held up fairly well.

The success of the Erie Canal inspired other states to construct canals designed to tap the reservoir of agricultural surplus beyond the Alleghany Mountains and to release manufactured goods east of the mountains. Between 1826 and 1834, Pennsylvania financed an interconnected system of canals and railroads which included 606 miles of canals and 118 miles of railroad by 1840. Unlike the Erie Canal, the Pennsylvania system did not pay for itself in tolls. But it benefitted farmers along the route, especially those west of Pittsburgh. Since Philadelphia was a principal terminal, the merchants and processors profited as handlers of agricultural commodities. The producers had a large and certain market for their excess output.

Maryland and Virginia hoped to get some of this farm trade by completing the Chesapeake and Ohio Canal, begun in 1828. But the canal did not reach Cumberland, Maryland, until 1850 because of engineering difficulties. By the time the builders completed it, the Baltimore and Ohio Railroad had entered the competition. Western states strove to link up with the Erie and other canals. Ohio, Indiana, and Michigan built canals linking the Ohio River with the Great Lakes and the Erie Canal. Produce which formerly flowed south to New Orleans gradually shifted northward to New York. By the 1840s, Buffalo handled more commerce than the port of New Orleans. This changed route of transport made commercial agriculture even more profitable along the newer arteries. The canals measurably helped shift the West from subsistence (or corn-hog) farming to grain growing.

Commerce on the Great Lakes served as a link between canals and markets. Those who wanted to reach into the interior had to improve Lakes shipping. In 1855, Michigan built the Sault Sainte Marie Canal, which joined Lake Superior and Lake Huron. The Federal Government improved the canal in the 1870s and took over its operation in 1880. Although the canal was used mostly to move iron ore and other industrial materials, it also handled large volumes of wheat and flour. Changes in ship transport reached the Great Lakes later. Steamboats were used on rivers and canals before their use on the Great Lakes.

BEGINNING OF THE STEAMBOAT ERA

Robert Fulton's *Clermont,* a 160-ton, side-wheel steamboat, made its first run on the Hudson River in 1807. Steamboating on American rivers had begun. In 1811, the *New Orleans,* built at Pittsburgh, made the first steamboat run down the Ohio and Mississippi rivers. At first steamboats only supplemented the major craft of Mississippi commerce, keelboats and flatboats. The steamboat became the most popular after design and engine improvements had been made. The keelboats, long used on western waters, differed little from such boats used since antiquity in the Old World. Boatmen poled keelboats upstream, and tremendous quantities of goods were thus slowly and expensively transported. It took 30 men some 90 days to push a keelboat from New Orleans to Cairo, Illinois. It cost a shipper more to make that trip than to move the same cargo frm Philadelphia to Pittsburgh by wagon. Shortly after the War of 1812, shallow draft river steamboats of huge capacity and great power had begun to dominate western waters. They continued to move farm commodities, people, and manufactured goods until after the Civil War.

By the mid-1840s, nearly 1,200 steamboats carried 10 million tons of freight on inland waters. The volume was twice that of American foreign commerce in any given year. By 1860, the whole vast Mississippi system had been reached by steamboats. Pittsburgh, Cincinnati, Louisville, St. Louis, Cairo, and New Orleans became great river ports. American river steamboat tonnage came to 868,000 tons by 1860. The figures cannot express the romance of *Life on the Mississippi,* but river steamboats were tremendously important in the development of a vast commercial agriculture in the interior.

Estimated Great Lakes tonnage rose from 7,000 tons in 1830 to 393,000 in 1860. Every type of goods moved on the Great Lakes, but most commodities were agricultural. Grain led the list. The first steamboat on the Great Lakes, *Walk-in-the-Water,* was launched in 1818. But steam shipping developed as slowly on the Great Lakes as on the oceans. By 1851, of the 215,000 tons of shipping on the Great Lakes, about one-third was moved by steam and the remainder by sail. Tonnage built for Great Lakes use rose from 403 in 1823 to 45,427 in 1855. The overall increase resulted almost entirely from the development of the canals. The numerical superiority of sailing ships on the Great Lakes belies their importance. The bigger, more reliable steamships made more runs and carried increasingly greater percentages of the Lakes freight.

The development of Lakes and river shipping made farmland near waterways useful for commercial agriculture. Land along the canals rose not just in price but in genuine value. The return cargoes included not only manufactured goods, but more pioneer farmers as well. By the 1830s, the Lakes traffic, which altered agriculture in the Old Northwest, also deflected the migration of farmers north of the moraine which marked the farthest advance of the Wisconsin glacier. The land north of the moraine consisted of soils impregnated with mechanically ground limestone. The immigrants began to move onto the most productive soil in the Midwest.

River steamboats made cotton even more profitable to growers in the Mississippi Delta and in the Black Belt of Alabama. The easy movement of food from the Old Northwest down the Mississippi made it possible for planters to concentrate on growing cotton instead of food crops. Despite these advantages, the water transportation systems had some disadvantages. Most of them froze in winter. More importantly, all methods of water transport were very slow. A faster means of moving goods and people was needed.

THE BUILDING OF THE RAILROADS

Recently scholars have argued about the importance of the railroads in the economic development of the United States. Like the ongoing dispute about the importance of the frontier in American history, the dispute over railroads will not likely be satisfactorily resolved. But clearly such a monumental undertaking had to have vast impact.

The building of the railroads, like the westward expansion, has no parallel in history. In 1830, the United States had 23 miles of operating railroad. This had risen to 1,098 miles by 1835, only five years later! The astounding increase continued: 4,633 miles in 1845, 18,374 by 1855, and 30,626 in 1860. The railroads unquestionably provided the fastest and cheapest form of land transport.

However, railroad freight charges exceeded canal, river, and Lakes shipping charges, except when railroads ran at a loss in order to crush or meet competition. Rail rates tended to be lower where the railroad faced water competition. Accurate average rail rates for the years before 1860 are not known. Canal rates averaged 2 cents a ton mile, river rates about 1 cent a ton mile, and rail rates around 4 cents a ton mile with water competition. Around 1850, rail rates sometimes went to 25 cents a ton mile without water competition. But rail transport still cost less than any other land transport.

By moving things faster the railroads saved money for the shippers in spite of the high rail rates. Railroads also ran in more nearly straight lines than did rivers. The distance between Cincinnati and St. Louis, for example, came to 327 miles by railroad, but it was 720 miles by way of the Ohio-Mississippi waterway. With a shorter route, the railroad could charge twice as much per ton mile and still be competitive.

Before the Civil War, however, the railroads did not provide continuous transportation for any great distance. Between 1850 and 1860, railroad operators tried to close these gaps, but many still remained when the Civil War began. Because the railroads did not all use the same gauges of track, cars of one line could not be run over the tracks of another. Through the mid-19th century, standardization of gauges gradually removed this difficulty.

At first Americans copied the British design of roadbeds, locomotives, and rolling stock. But the roadbeds proved too rigid, and the locomotives too heavy. American engineers had to make the roads more flexible and distribute the weight of the engines more evenly. Robert L. Stevens, an American, probably was the first to use wooden ties under the rails on the Camden and Amboy Railroad in the 1830s. In a land of abundant forests, this was a logical solution. John B. Jervis of the Mohawk and Hudson Railroad designed and built the first locomotive with a four-wheel truck attached to the engine by a swivel. This lengthened the engine, allowed distribution of weight over a larger surface, and still enabled the machine to get around curves. In 1837, Joseph Harrison invented the equalizing beam which spread the weight of the engine equally to all driving wheels.

As a result of these changes in locomotive design, American engines could take curves and grades which would have derailed most European engines. By the 1840s and 1850s, Americans actually exported locomotives to Europe. Although Americans never captured the market abroad, they made the superior engines until Europeans copied American technical advances.

The states had built the canals, but the age of railroads began during the worst times of the depression of 1837–1842. So private capitalists built most of the railroads with generous infusions of state and federal money. The people did not care to risk their own tax capital, so they left the chance of failure to private individuals. Since the people wanted the roads, the states gave some assistance, but with a good chance of recovering their loans.

The first state incorporation charters for the railroads either specifically regulated rates or, more frequently, made provisions for regulation. State legislatures issued the first charters since the first bureaus to handle incorporations did not appear until 1837. Even then entrepreneurs preferred to go to the legislatures because the railroads could get monopoly charters from the legislatures, but not from the bureaucrats who had no such authority. So special legislation created the corporations, and from the first, American railroads were at least technically subject to rate regulation. This later became an important issue, especially after the Civil War, but in the 1840s and 1850s it excited little public interest.

Railroads, even more than canals, influenced farmers in selecting their agricultural specialization. For example, dairying was developed along railroad lines. In 1851, cooled milk was moved from northern New York State to Boston. Grain could be sped from faraway Illinois, Wisconsin, and Michigan to be shipped to Liverpool and London. Manufactured products could be cheaply moved west, and the living standards of farmers improved. Farmers responded to the availability of cheap capital goods with increased commercial farming, made even more profitable with the building of railroads. But the new commercialism made the farmers dependent on the railroads.

The railroads as a group made up the first of the great corporations in America. Only later did manufacturers incorporate their companies. Corporations had the great advantages of being able to accumulate large quantities of capital and making it mobile. However, in the process of incorporation, actual ownership of the business became divorced from management. The railroads, owned and operated by corporations, were run by a professional managerial class who felt obligations only to themselves and their shareholders. They ignored the well-being of their patrons until abuses and oppressions had reached nearly intolerable proportions.

IMPACT OF RAILROADS ON FARMERS

The farmers' dependence on the railroads and the possibility of industrial tyranny resulted from the needs of commercial farming. But specialization would have been limited considerably had it not been for the railroads. The disadvantages of dependence on the railroads was temporary. Not only farmers but others benefitted from the development of railroads. Urbanites could obtain better and cheaper food.

Fast transportation by railroad had the most immediate influence on the producers of perishable commodities, the dairy and truck farmers. As early as the 1830s, some American cities had become so large and congested that milk could not be moved from farms to the center of the city and still be fresh. As a result, large cities, such as New York, developed urban dairies. Large cow barns sheltered cows fed on distillery and brewery slop, supplemented with garbage. The rows of stabled cows were milked with little regard for sanitation.

The railroad, aided by new health regulations, slowly drove the urban dairies out of business. During 1842–1843, the Erie Railroad carried more than 750,000 gallons of milk into New York City. These deliveries increased greatly during the next decade in New York and other large cities as well.

The milk was refrigerated after a fashion by stirring it in cans with ice-filled tubes. The milk was cooled enough to reach the market in acceptable condition. Dairymen formed associations, set up receiving stations in the cities, and ran city delivery services. The consumer not only received a comparatively fresher product, with fewer health hazards, but the milk cost less. As a result, the per capita consumption of milk in the cities rapidly increased. This stimulated a greater specialization in fluid milk production on dairy farms near cities. As the railroads expanded, so did commercial dairying, which soon became one of the most significant farm specializations.

Producers of fresh fruits and vegetables also benefitted from the railroad. Previously truck farmers were limited to the immediate vicinity of cities. The railroads, especially in New York and New Jersey, the "Garden State," caused an increase in sources of supply during the 1840s and 1850s. Special express trains rushed truck produce to the city. Farmers began to grow many horticultural specialties, especially strawberries. The great increase in the supply of these specialties did not cause prices to fall because as the production expanded so did the demand. Urban consumers had long wanted the commodities now made accessible by the railroad.

The producers of dairy and truck crops became the first to depend on the railroads and also became the first victims of railroad policies. The truck farmer simply had to use the railroads, regardless of freight rates, because he could not store his crop. A western wheat farmer could at least in theory hold his product. Some milk could be used for cheese, but even this was a limited method for preserving milk. So, although not widely known, the eastern farmers were the first to object to rail monopolies. On the other hand, the proliferation of rail lines in the East sometimes provided the farmers with alternate routes, and on the coast, short distance water transportation sometimes kept rail rates down.

The supply of certain fruits and vegetables became more certain and more continuous with the advance of railroads into the South. In the 1850s, New York City obtained fruits and vegetables from Virginia, the Carolinas, and Georgia. Truck farms and orchards flourished, particularly in the vicinity of shipping points such as Norfolk, Charleston, and Richmond. Some southern rail lines ran to the interior, and fruits were carried to ports and shipped by water to New York.

The cities of the interior and the South experienced similar changes. Fruit and vegetable growing and dairying were developed to supply Chicago, St. Louis, Cincinnati, and other cities. The rise of Wisconsin as a prominent dairying area can be linked to the growth of Chicago. Chicago merchants and railroads also reached into the Deep South for various horticultural specialties, a very significant development.

In the transport of perishable commodities, speed was essential because good refrigeration had not been developed. Chemical preservation and pasteurization were unknown or little used. Subsequent advances in bacteriology and in refrigeration should not obscure the great importance of fast transport of perishable commodities during the earlier stages of American commercial farming in the mid-19th century.

CHAPTER NINE

GROWING THE CROPS ✕ 1783–1865

F EW TASKS are so difficult and time-consuming as plowing, harrowing, and cultivating the soil. The plow has long symbolized agriculture, and the plow and its power source have long determined
the shapes of fields and types of soil used. Indeed, most other agricultural machines or devices became economically useful as a result of
advances in plows and plowing techniques because the preparation of
the soil took precedence over other farm tasks. In 1790, about 90
percent of the population were farmers; this percentage declined to
72 percent in 1820, 69 percent in 1840, and 60 percent in 1860. Great
scientific and technological advances were made in the handling of
crops during those years.

IMPROVEMENTS IN PLOWS

The first significant advances in plow making resulted from efforts to
rationalize and standardize plow design by the application of scientific
or mathematical principles to the design and construction of plows.
The first American credited with such an effort was Thomas Jefferson.
He hoped to establish a uniform design which could be used by plow
makers everywhere. Jefferson did not realize this dream but others
did. Uniformity allowed the application of industrialization to plow
manufacture. Thus plow making benefitted from the savings which
accompany a shift from craftsman to factory production.

Until the 18th century, Americans made their plows chiefly of
wood and used as little as possible of the comparatively scarce and
expensive iron and steel. However, the cutting and wearing parts
were usually iron, with iron straps over the moldboard. The increasing
iron technology and production made all-metal plows a possibility.

In 1797, Charles Newbold of Burlington, New Jersey, secured the

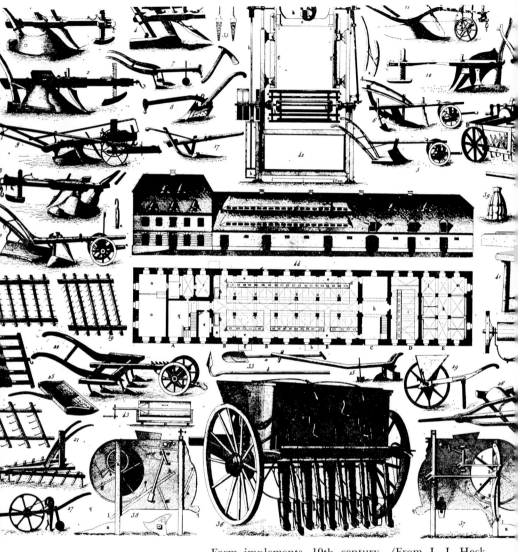

Farm implements, 19th century. (From J. J. Heck, *Iconographic Encyclopaedia*, 1851) Courtesy Library of Congress

first American patent for a cast-iron plow. He cast the whole plow—share, moldboard, and landside—in one piece. The disadvantage was that if any part of the plow broke it could not be repaired. Newbold tried unsuccessfully to develop a large market for his plow. Farmers rejected it probably because its cost was too high in relation to its effective life. Even if the plow did not break, the share dulled rapidly and the farmer had to replace the entire plow.

In 1813, R. B. Chenaworth of Baltimore patented a cast-iron plow with a separate share, moldboard, and landside. As the parts of the plow wore out or broke, the owner could replace them at a moderate cost. However, credit for this invention has commonly gone to Jethro Wood of Scipio, New York, who patented his plow improvements in 1814 and 1819. He must have known of Chenaworth's invention, for Wood did not claim that he had invented replaceable parts. He made his application for a new method of joining the parts without using screws or bolts, which were difficult to make and expensive. Jethro Wood also apparently improved the design of the moldboard. He publicized his plows and introduced easily interchanged, uniform parts. Wood manufactured his plows quite successfully on a large scale.

Other technolognical advances occurred which had to be integrated by plow manufacturers before the improved plows came into wide use. In 1817, Edwin Stevens of New Jersey devised a method of cold chilling the cutting parts of the iron plow which increased its wearing quality. During the next twenty years, inventors made many improvements in plows, including better moldboard designs for different qualities of soil. Plows with reversible or revolving moldboards and shares were "reinvented" for use on hillsides, so that the furrows could always be thrown in the same direction. Although Europeans had devised sidehill plows during the Renaissance, these plows were unknown in America. As the plows were improved and as mass production made them cheaper, farmers in the East and parts of the Midwest gradually shifted from wooden to iron plows. Between 1830 and 1845, from Massachusetts to Indiana and from Maine to Alabama, the iron plow was adopted nearly everywhere. The iron plow started a farming revolution.

Changes in the South followed patterns slightly different from those in the North. In parts of the South, slaves still used hoes to prepare the land. This seemed to be commonest in the lowland coastal plains of the Carolinas and Georgia. Between 1783 and 1865, southerners mostly used the shovel plow, which northerners had commonly used during the era of the wooden plow. Some northern and southern farmers still used shovel plows until after the Civil War. The shovel plow was essentially a spade with a loop on the back for the stock or standard to enter. The handles and beam were attached to the stock. The plow cast the dirt two ways, cut a shallow furrow, and did not turn the soil as much as the moldboard plow. Although generally considered an inferior plow, it actually caused less erosion than other kinds. The southern farmers had adopted an iron plow before the northerners. The shovel plow limited farmers to using land which could be successfully cultivated with a light draft and shallow furrow. On the other hand, southern hoe agriculture required vast amounts of slave labor.

As wooden plows gave way to iron in the North, so the ancient shovel plow gave way to the Carey or Dragon plow in the South. These basically cast-iron plows had interchangeable and replaceable parts, but were generally lighter weight than plows made for use in the North. The southern plows increasingly were made in the North by plow manufacturers instead of by plantation blacksmiths. The moldboard plows of the South were used mostly in Virginia and in the Piedmont areas. Some plows of the stronger wrought iron were used in Alabama and Mississippi in the 1850s. For the most part, farmers and planters did not use cast-iron moldboard plows until the 1840s and later.

The shovel plows often could not break sod, so farmers might use a hoe and axe. Then they planted corn whose roots further broke the soil. They would then be able to plow the next year. Some planters did use wooden plows as breaking plows, as also happened in the North. Southerners also used trowel-hoe plows as breaking plows; these were single shares pulled through the dirt to cut a narrow trench, not a real furrow. The shovel plow, used in broken soil, cut about three inches deep, about the depth reached by the ancient Roman plows.

EFFECTS OF IMPROVED PLOWS ON FARMING

On the whole, farmers did not need large furrows and deep cuts. Mostly, the improved plows reduced the amount of man and animal labor required to work the soil. These savings did not count for so much in the South where planters kept slaves for cultivating and harvesting the crops. The shovel plow typically required only one mule, and even a girl could handle it easily. Southern opposition to the moldboard plows probably derived from the fact that they required more animal power and skilled plowmen.

In the North, however, the newer cast-iron plows markedly reduced the man and animal power needed. If the plowman used oxen, the number of animals needed fell from two yoke to one yoke. In the case of horses, need fell from three horses to two. A reduction in the number of draft animals required meant considerable savings for the

Old Colony strong plow, New England, early 19th century.

ordinary northern farmer. The iron plows also could be used by one man instead of two or three. The amount of work which a farmer could do in a day increased by 50 percent to 100 percent, or from an acre a day to an acre and a half, or perhaps two acres a day. This saving should be calculated as the amount which the plowman could accomplish in a plowing season.

The farmer could theoretically double his production if he doubled the amount plowed, and the size of farms did increase. However, farmers had an important limitation in the amount they could harvest without mechanical reapers, rather than in the amount they could plow.

Plow makers varied their plows to suit different soils and uses, mainly by changing the curvature of the moldboards, altering coulters, and sometimes adding wheels to the beam. Sod, clay, sand, and stubble—all had their special plows. Cultivating plows and gang plows were introduced. These developments and variations proved very successful in the Northeast, less so in the Ohio valley and Lakes plains, and even less so on the prairies of the Midwest and the West. The deep-rooted sod and the sticky lime-rich soil presented two difficult problems for the early prairie farmers.

BREAKING THE PRAIRIE

First, the cast-iron plow could hardly cut through the heavy sod, and prairie farmers had to replace both shares and coulters frequently. The wooden plow usually had a wrought iron share, hammered out at a blacksmith's forge. These shares wore well and could be sharpened. However, in breaking the land the soil made the plows hard to pull, and the farmer lost any advantages of less draft power. If he had to use a large number of animals in breaking the prairie sod, the wooden plow served the pioneer better since it took no more animals or men and cut the sod quickly and properly. From the 1820s through the 1840s, the most useful and popular sodbusting plow was made of wood with wrought iron share and coulter and a wooden moldboard covered with iron strips. It ran heavy, but it did the job.

The whole outfit was very heavy, which had the advantage of holding it down. The plow alone, without handles or beams, often weighed 125 pounds. The fourteen- or fifteen-foot beam and handles added more weight. The total weight was supported at the front of the beam by two small wooden wheels which regulated the depth and width of the cut. Such a prairie sod breaker required three to seven yoke of oxen for draft, a plowman, and possibly a boy to prod the oxen. Perhaps eight acres could be worked in a plowing season. Sodbusting was often custom done, so the plowman might continue to plow after planting was finished on other parts of the field.

The conquest of the soil was indeed time-consuming on the prairies. The settlers first cleared and plowed the forested land which could be worked more easily. Pioneers near prairie would plow about three rows into the surrounding grassland as a firebreak. The next year the farmer would plant this and would plow a new firebreak. As farmers advanced onto the prairie and discovered its fertility, they devised and employed the special prairie sod breaker to speed the advance.

Putting in the sod crop, or the first corn crop, on the newly broken soils called for a second plowing the year after the breaking. Cast-iron plows were used for this, which brought out the second problem. The decayed and broken sod accumulated on the plow until progress became impossible. The plowman had to carry a paddle and scrape off the plow every few feet. Some speculated the prairie lands would have to be abandoned because of the difficulty of the plowing after the sod breaking plowing. The wooden plow was far too slow and inefficient.

IMPROVED PLOW MATERIALS

John Lane, a blacksmith of Lockport, Illinois, is credited with the first successful effort to decrease the clogging on the second plowing. In 1833, Lane made a moldboard with a highly polished surface. The plow scoured well and the sticky soil slid off. He also made strips of saw-blade steel which he placed over the wooden moldboards of the prairie plows. Lane's innovations made the cultivation of the prairie sod a practical possibility.

Reconstruction (based on 1838 plow) of John Deere
steel plow, 1837.

Lane did not patent his idea, and he did not manufacture his
plows extensively. The steel he used was made by the old method in
small ovens, and it was both scarce and expensive. Steel continued to
be impractical until after the development of the Bessemer process
and the open hearth methods of manufacture. Extensive use of steel
was unnecessary, and plowmakers did not use much steel until improved technology brought the price down.

In 1837, John Deere began making simple, one-piece plows of
wrought iron, with a cutting edge of steel on the share. The steel was
carefully welded to the moldboard. Plowshare and moldboard designs
were important in the application of the method, for the plow
amounted to a curved trapezoid which permitted the successful merg-
ing of iron and steel. Deere's plow scoured so well that he and others
described it as the "singing plow." Deere sold some plows during the
next ten years, but not in any great volume. He gradually refined the
plow design until it looked more like the traditional plow. It still combined the wrought iron moldboard with the steel share.

In 1846, Deere and his partner, Leonard Andrus, pushed production to about a thousand plows a year, and in 1847, the Deere plant
was moved from Grand Detour to Moline, Illinois. Deere made ten
thousand plows a year by 1857, and other manufacturers turned out
similar plows. As cheaper steel became available in the 1850s, all
manufacturers used more of the comparatively expensive metal. But
the wrought iron-and-steel plow had caught on with the farmers of
the West and even the East.

USE OF GANG PLOWS

Plows could actually be run too fast through the soil. Plows, designed to be pulled by one or two horses through heavy soil, did
not work well when pulled more rapidly by the addition of greater

draft power (as in the use of tractors) or by the reduction of the drag in loose types of soil. Generally, the plows worked well at the speed of a walking horse, or about five miles per hour. Farmers usually accommodated to the peculiarity of a particular plow by maintaining a certain speed. The plowman could slow everything down by adding plows or by using gang plows.

On the other hand, plows designed for fairly rapid movement through the soil did not work well when pulled too slowly. The simplest way to increase productivity was to increase the number of plows rather than alter the speed of plowing. Thus gang plows, with sometimes as many as 12 or 14 plows, came to be used with many horses or with steam tractors. Gang plows required a lighter drag of the iron plows and the setting of the cut to shallower depths.

Until the Civil War, gang plows were used only in the North where the relatively short plowing season and the shortage of labor made expeditious plowing necessary. The southern small farmer had limited land to plow. His ability to harvest the crop also regulated the amount of land he could use. Because he could easily plow more than he could harvest in cotton, he had no need for a faster plow. The invention of machines for harvesting hay and grain crops enabled the farmer to grow more per man, and thus he could plow more per man. Consequently, plow improvements meant more to the northern farmer until the invention of cotton pickers and a shift in crop specialties produced a need for plowing efficiency in the South.

PREPARING THE SOIL FOR PLANTING

The animal-drawn moldboard plow broke the soil, but seldom enough for immediate planting. The slowly drawn plow left large chunks of soil and tended to produce a corrugated pattern of soil deposit. After plowing, the farmer harrowed the field to further prepare it for planting. Harrows had various forms, almost all of simple design and function. The triangular or "A" harrow had been used since the 17th century in America, and it continued to be used through the Civil War, especially on rocky or badly cleared fields. Southerners, particularly cotton growers, usually relied on cross plowing with the shovel plow. Because it cast such a light furrow, the shovel plow broke up the soil without turning over large clods.

In the 1840s, a two-horse, hinged harrow appeared which proved more economical and effective on well-cleared land. This harrow was essentially a hinged trapezoid with iron or steel spikes. If the spikes caught on rocks or roots, they were often broken, or it slowed the harrow. In 1869, the spring tooth harrow appeared. The spikes of this

Joseph Sutter's Patent Gang Plow, 1859. (From *American Agriculturist,* Vol. 18, Oct. 1859)

harrow were attached to a spring so that when a spike hit an obstacle it lifted up instead of breaking off.

In the North and West, farmers sometimes used gang shovel plows to prepare the seedbed after plowing. But farmers used other implements for this work as well. By the 1840s (and possibly in the 1830s), field cultivators were substituted for harrows and light plows. These probably appeared first in New York where farmers used them instead of the harrow for covering the seed. Farmers also used the cultivators to prepare the seedbed.

The field cultivator had a rectangular frame that had several very small shovel plows attached to it, which broke the soil or covered the seed. The cultivator was easy to draw and worked fast, although it covered the seed more efficiently than it broke up the seedbed. Such cultivators had nearly disappeared by the Civil War. The field cultivator helped cross plowing in the Northeast, particularly in New York.

Either before or after seeding, the farmer might work the field with a roller. The moldboard plow (seldom used in the South) and the nature of the soil largely determined whether or not the farmer used a roller. The roller worked well where farmers used the moldboard plow. It reduced the ridges of the furrows and cut down the smaller clods. Farmers generally made their rollers from large logs set in axles and pulled by one or two horses. They set rocks as weights on a frame to press the rollers down if the logs were too lightweight.

Design improvements resulted in tillage implements which had less drag, cleared obstructions better, and moved more rapidly and effectively across a field. Saving time became more important as the amount of land in crops increased.

PLANTING THE CROPS

Planting or seeding also involved some movement of the soil. Different crops required different seeding techniques, and farmers tried to substitute mechanical seeding devices to save labor and seeds and to secure better stands of crops. In many crops, the precision of seed placement aided later cultivation or gave the plants space for growth and a sufficient area from which to draw nutrients and water.

Planting, whether for large seed crops such as corn and cotton or quarters of potatoes, followed a standard form. Farmers widened the plowed furrows with a hoe or hoelike implement. They then deposited the seed, or seeds, in the hole and closed it. To plant corn, one worker hoed and the other dropped four or five seeds in the hole and covered it with his heel. Cotton planters usually used three slaves: one to hoe, one to drop the seed, and one to rake soil over the seed.

Generally, smaller seed was broadcast by hand. Grass, clover, wheat, rye, rice, and even tobacco were spread by hand across the seedbed. The plants were sometimes thinned, as in the case of carrots, or transplanted, as was often true with rice and tobacco. The small tobacco seeds were usually mixed with sand in order to get a better seed distribution in the planting. The seedlings were later transplanted. Devices for placing seeds uniformly, called drills, were used in England during the 18th century. Drills were introduced during the Revolution, but they were not widely used. Some Americans used drills for wheat and other small grains; but if so, their drills were imported from England or were homemade.

Dickey's Improved Patented Corn Planter, 1857. (From *American Agriculturist*, Vol. 16, New York, Apr. 1857)

Randall Jones' two-row hand corn planter, 1850s. (From P. W. Bidwell and J. I. Falconer, *History of Agriculture in the Northern United States, 1620–1860*, Washington, D.C., 1925)

Corn and cotton had large seeds which suited them to mechanical planters. A measured amount of seed had to be regularly spaced at comparatively wide intervals. Methods of opening a place for the seed, for covering it, and for regulating the depth of planting had to be developed. Corn culture also required hilling, which required a checkerboard planting, especially if the planter used horse-drawn cultivators. These requirements delayed the use of drills for planting larger seeds.

The use of grain drills was impeded by the many stumps and rocks in the fields. Drills lost effectiveness if they broke or were jolted out of the ground. Drills were first used in long-cultivated areas and later in the comparatively stump- and rock-free prairies.

Commercial production of seed drills began around 1841 and had not increased much by 1860. In the 1840s and 1850s, the grain drill appeared more frequently in New York and Pennsylvania, and later spread to Delaware, New Jersey, and the Midwest. By the late 1850s, it had become commonplace in the Middle Atlantic states. Although common in the Midwest, it was seldom used west of Illinois and Wisconsin.

The first drills did not measure the grain exactly. The drills could deposit either too much or too little seed if not carefully watched and regulated. Two general methods of depositing seed were used. One

Grain drill planter, *ca.* 1850s.

device used a perforated revolving cylinder which released the seed directly into the soil, but the seed could be conveyed to the ground through hollow stems. The soil either fell back onto the seed or, more often, was pushed back by small wedges trailing behind the stems.

The best drills used a box with some sort of vibrating mechanism which kept the seed moving through hollow stems. Regulating the flow of seed was improved by using cams to trip a flap covering the hole through which the seed fell. Drills often clogged or spread the seed unevenly. The farmer had to plow and harrow his soil very well before the machine would work properly.

Western farmers usually broadcast small grains on former corn or tobacco fields. Shallow and incomplete plowing prevented the early drills from working properly. Although "shotgunning" did not provide an adequate seed distribution, the method was used to seed wheatlands which were not carefully plowed. Such an inefficient seeding method was commoner in the West than in the East. Westerners preferred broadcast seeders over drills. Some were hand operated, and some were powered from ground wheels, which sprinkled or blew the seed quite evenly over the ground. Harrows then covered the seed, at the same time or later.

The seeders allowed considerable savings in seed. Better germination rates usually resulted from the use of seeders, especially drills, thus helping to improve the seed to yield ratio. Farmers could get as much as eight bushels an acre more than the yields from hand broadcasting. Farmers saved even more in man-hours, and mechanized broadcast seeders saved as much time as drills.

Changes in seeding methods reduced labor costs and also allowed increased production profitably at a greater distance from the urban market. In addition, the horse-powered drills and seeders proved most profitable on the larger acreages; on smaller farms, hand-powered seeders did the job as well and cost far less. Typically, this technological advance most benefitted the large operator and encouraged large-scale farming, and it actually put the small farmer at a competitive disadvantage.

Until the 1850s, farmers had to plant corn by hand. Hollow canes, which deposited seeds in measured amounts when pushed into the ground or when hand-triggered, were invented; but the seed still had to be covered with a hoe. Two hollow canes mounted on a crosspiece allowed the planter to plant two rows at once. For the most part, the hand corn planters were popular only for planting sweet corn in gardens. In small patches the speed of planting was worth the hard work necessary to use the planter.

Horse-drawn corn planters attracted inventive efforts for many years after the small-grain drill appeared. Both corn and cotton planters needed a device that would space the seed, and the corn producer who wanted to cultivate both ways in a field needed a machine that would plant in check rows. In 1853, George W. Brown of Galesburg, Illinois, received his first patent on a corn planter from which the seed was released by levers operated by the driver, or more often, another person. The delivery through hollow stems, the conveying of

Haworth corn planter, patented 1861.

the seed by shovels, and pressing down by rollers readapted devices used for seeding small grains. Brown built up a good business in the 1850s, and other inventors also entered the field successfully.

The farmer cross-marked his fields in the checkerboard pattern with a marker, so the lever was pulled at the right places. Some planters required the driver to push a pedal to release the seed. The next development was to run a wire or cord with regularly spaced knots down the field. This tripped the seeding mechanism. These devices, which appeared around 1865, did not work very well.

Cotton was usually planted in hilled rows by slaves, although mechanical cotton planters came into use in the 1840s (about the same time as corn planters). A homemade hollow drum with widely spaced holes was most often used, and the seed was not carefully measured until after the Civil War. A hoe, shovel plow, or cultivator was used to cover the seed.

CULTIVATING THE LAND

After the crops were planted, the eternal battle against weeds began. The ancient method of killing weeds was to chop them out with a hoe. Gardeners had long used it in corn and cotton fields and in vegetable plots. Small farmers and slaves also used the hoe to weed and thin sugar and cotton plants. This hoeing was commonly called chopping. The hoe continued as the main cultivating tool, although planters sometimes used the shovel plow to cultivate between rows of cotton and other crops.

In the North, including the Old Northwest, farmers hoed garden crops such as potatoes. They also used the small plow, either shovel or cultivating type, to aerate and weed their field crops. About 1820, specialized, cheap cultivators appeared, which weighed much less than plows and supplanted the hoe and small plow by around 1840. One cultivator could till more land than three plows in any given time. The cultivator had a triangular frame with handles and a clevis, which was used for attaching a horse harness. The frame had small feet (like triangular spikes and narrower than shovels) set into it.

In the 1830s, the expanding cultivator appeared. Its frame was constructed so that it could be made wider or narrower in order to fit differently spaced rows. Only one row could be cultivated at a time with this cultivator. The cast-iron teeth, fastened in with a short neck, caught on stumps, rocks, or roots; and they were harder to clear of these obstructions than a simple spike. Nevertheless, the device was an important advance for the increasing of cultivated areas.

By 1840, farmers in the East no longer used plows to cultivate

Picking cotton, 1850s. Wood engraving from *Harper's New Monthly Magazine,* March 1854. Courtesy Library of Congress

corn. In the Old Northwest, farmers still commonly used a light, one-horse shovel plow. The number of shovels was increased to three, and steel replaced iron as it did in plows and other implements.

Sometime in the 1840s, the sulky (or riding), straddle-row cultivator came into use in cornfields. One horse and one wheel went down each side of the row, and shovels or teeth cut close to the corn plants. This cultivator doubled the amount of corn a farmer could conveniently cultivate. The riding straddle-row cultivator appeared first in the East; but by the 1860s, it became widespread in the prairie and corn country.

In 1856, George Easterly of Wisconsin took out the first patent on this cultivator, although he had not invented it. Soon other farmers and inventors made changes and improvements. Attachments were added to protect plants, to raise and lower the shovels, and to allow the shovels to rise over obstructions. Another type, the two-row walking cultivator, also found favor. It had no tongue, so it was easier to turn at the end of a row.

The new cultivators not only saved man-hours, but they did the work more thoroughly. Cultivators were also used in the South with row crop grains. In the 1840s, cotton growers cultivated with the same implements used in the North for corn. Cotton farmers sometimes cultivated with the sweep, a sharpened piece of wrought iron fastened to a wooden plank, set a little higher at the rear than at the front.

Drawn between the rows of cotton, it operated much like a razor to shave off the weeds. Varieties of plows were also used to cultivate cotton, to keep down the weeds, and to throw small mounds of earth up onto the plants. But most planters still relied on slaves with hoes to do most of the cultivating. If slaves were not used to hoe, they might be leased to town craftsmen.

Southwestern farmers, who had access to the food of the Old Northwest, could concentrate on increasing cotton production, and they used slaves less to produce food or do other work. By most accounts, southwestern planters also worked their slaves harder to increase production. On the whole, western cotton producers made more and better use of plows, harrows, cultivators, seed drills, and other implements and in so doing grew more at a lower cost than planters in the East. But a key factor still was successfully judging the amount that could be harvested by the labor force on hand, and this determined the production and thereby the prosperity of the farmer.

Between 1783 and 1865, soil-working implements underwent improvements in design, in materials used, and in methods of manufacture. Interchangeability of parts and mass factory production brought economies, which produced benefits for farmers, and also forced standardization of styles, designs, and materials. New problems presented by the prairie soils required solutions with all kinds of new implements. Most of the inventions and discoveries were made by men with some farm experience who were no longer farmers. Inventions for the farm resulted most often from inventions off the farm, as in the case of steel in farm implements. Special solutions for particular problems rapidly became general solutions.

CHAPTER TEN

HARVESTING AND PROCESSING ✖ 1783–1865

F ARMERS must do everything in season. For the farmer, the production process always starts several months before it ends, and he can do little to speed the biological processes which take place between its beginning and end. Furthermore, advances in soil preparation and seeding cannot help the farmer much if harvesting does not keep pace. Consequently, production invariably hinges on how much the farmers can get done in each season.

As the Revolutionary War ended, the cradle for harvesting grain was introduced in the Middle Atlantic states. In other places, ancient methods predominated until the early 19th century. Invented in Europe, the cradle consisted of a scythe attached to a frame with four or five long fingers, conforming to the curve of the scythe and extending above and slightly backward from the blade. The cut grain fell onto the cradle fingers; and with a turn to the side at the completion of the cut, the reaper deposited the grain. It could then be raked and bound into sheaves. "Throwing" the cradle required experience, skill, and strength. However, those who mastered the art could cut more than twice as much as those who used a sickle, reaping hook, or scythe. The sickle was used to cut grain fairly close to the heads. The scythe was used mostly for cutting grass and hay. It caused grain to shatter.

The amount of grain or hay that a harvester could cut in a day depended on the denseness of the stand. He could cut perhaps half an acre in a field of thick grass or grain, using a sickle or a scythe. In a lighter stand of grain or grass, reaping or mowing would average three-fourths to one acre. But, with the cradle, an expert could reap two or three acres of grain.

Around 1820, an improved version of the cradle was made by Charles Vaughn of Hallowell, Maine. The better design and lighter weight of the Vaughn cradle reduced the work of harvesting. Vaughan may have derived his ideas from similar cradles used in Scotland. In any case, farmers often called his cradle the "Scotch Bow" or the Scotch cradle. Cradling ceased to be a highly specialized skill. Not only could the farmer save the higher wages of cradlers, but he could also cut more grain in a day. The new cradle spread to the West from 1820 to 1840 and persisted in most grain areas well into the 1860s, even after mechanical reapers had appeared.

DEVELOPMENT OF MECHANICAL REAPERS

The story of mechanical reapers is tangled, and the historical difficulties arise from two main causes. First, inventors and manufacturers made large profits from selling the reapers, and this led to patent fights and advertising assertions of priority of invention. Inventors sought to ignore, denigrate, or minimize the contributions of others, but a pride in being the first may have stimulated many inventors to assert their contributions. Even the mechanic (and possibly also blacksmith) Jo Anderson, who worked for Cyrus McCormick, has been the subject of praise and fame, largely because he happened to be black.

The second difficulty in tracing reaper history arises from the "Great Man" theory of history. This inspiring theory holds out the prospect that those ignored now may yet be remembered by posterity. If seminal minds, rather than the inexorable movement of science and technology, account for inventions, then it matters greatly who did what. (It makes little difference to those who believe that sooner or later someone would do it anyway.) Disputes center around priorities and personalities. Discussion of the invention of the reaper sometimes opens with a notation of the shortcomings of the early machines and of their several inventors. However, the discovery that Hussey's machine had a badly balanced draft or that McCormick's cutter bar proved virtually useless should not come as a surprise to anyone. Despite these shortcomings, the ideas of both men and many others gradually merged to produce an economically successful machine.

Obed Hussey invented and marketed the first commercially successful reaper. Cyrus McCormick invented several parts of the reaper which proved so successful that his version eventually supplanted Hussey's. The most effective cutter bar, which is still used on reapers, was invented by Hussey. McCormick secured the patent rights to the Hussey cutter bar around 1850. A number of manufacturers sold these rather inadequate machines from 1833 to 1850. However, inventors had added seats for the machine operators, a mechanical reel to push the grain into the cutter bar, and rakes to remove the cut

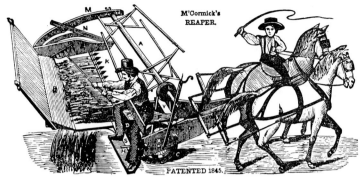

M'CORMICK'S REAPER.

This machine was patented in 1845, by C. H. M'Cormick, of ————, Virginia. It has been extensively used in most of the grain growing States of the Union, and if we may judge from the numerous certificates of those by whom it has been tried, it is a very effective and valuable implement. By reference to Mr. M'C's advertisement, to be found in this number, it will be seen that the machine is warranted to cut from fifteen to twenty acres of grain in a day, and at a great saving of expense over the common mode of harvesting.

Advertisement for McCormick reaper, patented 1845.

grain from the platform before 1860. The Civil War stimulated use of reapers by causing both farm prosperity and a labor shortage.

McCormick's early machines (before 1854) did not operate very well as mowers. The cutter bar of a mower has to operate swiftly and sharply through the thick stands of grasses and legumes. But less efficient cutting did not seriously hamper the effectiveness of a reaper. Farmers, however, wanted a machine which could be used both for reaping and mowing.

McCormick's reaper was improved when he adopted two inventions of Hussey: the cutter sickle in 1850 and the open back guard in 1854. The sales of the McComick machines shot up in 1855, 1856, and 1857. The McCormick bookkeeper of the time noted that those machines were the first to be generally satisfactory. But a worker still had to sit on the machine and rake grain or fodder off the platform.

SELF-RAKING REAPERS

In 1854, at least four firms began to manufacture self-raking reapers in places as far apart as New York and Illinois. Two more firms entered the field in 1855, but only the Atkins Automation Reaper, built at Chicago, was of any importance. The leading makes of hand-rake reapers (such as the McCormick, the Manny, and the

Wood) continued to dominate the market. In the wheat areas of Illinois, Wisconsin, and even New York, the self-raking reapers found large markets even before the Civil War. Continual improvements resulted in very effective self-raking mechanisms about the same time the Civil War caused a labor shortage. With the sudden rise in grain prices during the war, farmers could afford the new self-rakers and the hand-raker virtually disappeared.

No matter how the grain was raked, someone still had to follow along behind to hand rake and bind the grain for shocking in the field. Depending on the size and efficiency of the reaper, three, six, or more men followed closely behind so that the grain already cut was not trampled on the return run. The effectiveness of the reaper depended on the number of men who worked behind the machine. The reapers were designed for use in humid areas, where farmers had to dry the grain before threshing.

INVENTION OF REAPER-HARVESTERS

In 1836, Hiram Moore and J. Hascall of Kalamazoo, Michigan, made the first moderately successful machine that cut and threshed grain in one operation in the field. Their combine, which also cleaned and bagged the grain, had to be modified from the patent specifications, but the machine worked. It had a reciprocating sickle bar, a reel for pushing the grain into the cutters, and a conveyor for carrying the grain to the threshing and cleaning machinery. The machinery was powered by two ground wheels: one ran the reaping machinery and the other ran the threshing machinery.

The inventors made several of these machines, and one was still used in Michigan as late as 1843. Its owner worked it on custom and could harvest 25 acres of wheat in a day. In 1853, he sent the machine to California where it still performed well and inspired many copies or modifications by the wheat farmers of California, particularly in the San Joaquin valley. In the Midwest, however, the McCormick reaper gradually replaced the few combines which had been used because the humid climate made shocking and threshing the wheat preferable.

In the Midwest, inventors and farmers directed their attention to machines which would reduce the labor and the time needed for gathering and binding the cut grain. In Illinois, around 1851, an experimental reaping machine carried men on the machine to bind the grain. Farmers tried out other reaper-harvesters, most of which used a moving apron to deliver the grain to the binders. In 1858, C. W.

Hiram Moore (in high hat) on his combine, *ca.* 1850.
Photograph from collection of F. Hal Higgins

Marsh and W. W. Marsh, farmers of DeKalb County, Illinois, invented and successfully operated a harvester, known as the Marsh
harvester. After an apron delivered the grain to a box, two men riding on the machine bound it. The Marsh brothers continued to
build machines, and they worked out the last kinks in 1864. The
Marsh harvester went into successful production, and the Marsh
brothers had made about 50 of the machines by 1865.

The best solution was a reaper which automatically bound the
grain. Experiments with binders and binding materials started by
1850, but none of the early self-binders worked very well. Success
came only after 1865.

Production per man-hour in reaping and harvesting increased
steadily from the 1830s. Cradlers averaged two acres a day; with the
Hussey reaper, the cut rose to around 15 acres a day. McCormick's first
successful reaper (1855) handled 20 acres a day under ideal conditions.
But the machine had only a five-foot cut, and it could not reap much
more than 15 acres a day under ordinary conditions. Twelve acres
average was a good cut in 1855, regardless of the reaper used. Roughly,
one reaper equaled five or six cradlers.

Prior to the self-rake reaper, one man had to rake, and usually
two binders followed the reaper. The self-raker probably saved the
labor of five men. The Marsh harvester reduced the labor crew by

another two or three men, leaving three at the machine. Farmers using mechanized reapers could reduce the size of the harvest crew and still finish the work more quickly. Using the harvester, less grain shattered during the raking and binding. Although he had to keep horses, the expense of keeping horses came to far less than the wages of a harvest crew.

Small grains had to be threshed and then winnowed or separated. Until about 1825, farmers either beat the grain with flails or had animals walk on it. Horses and mules did the treading faster and more efficiently than cattle, and animals performed more efficiently than men with flails. Threshing by animal power cost less in the long run than using human labor. The commercial grain farmers in New York, Pennsylvania, and the West used treading from the end of the Revolution through the 1830s, when threshing machines became fairly common. Farmers with acreages of wheat sufficient to use mechanical reapers used their horses for treading.

Generally speaking, the New England farmer used the flail. On his typically small acreage, he planted grain for family use rather than for sale. In most of the Middle Atlantic states and the South, treading predominated. Southern farmers and planters seldom used platforms but had the grain trodden out on the ground. Commercial grain farmers of Pennsylvania and New York usually used treading platforms. Platforms gradually became more widely used in the South, particularly in Maryland and Virginia.

Treading on the ground produced dirty grain, which brought a lower price at the mill and a lower price as flour. Since wheat was seldom an important southern cash crop, it was not worth expensive, intensive methods of growing or harvesting. Grain produced for local markets, as most southern grain was, did not require the careful attention given a commodity intended for the international market.

DEVELOPMENT OF THRESHING MACHINES

Threshing machines appeared in the United States in the early 19th century. Most early machines copied the flailing method. In threshers (as in other machines), some time passed before men substituted a rotary action for the reciprocal action of the human arm. Around 1820, inventors devised cylindrical threshers with spikes. Farmers in the Middle Atlantic wheat regions used them some, although the flail and the horse remained the chief method of threshing. The rapid expansion of the thresher, either man- or animal-powered, took place in the 1830s.

By the mid-1830s, about 700 different types of threshing machines were being sold, although most of them were shoddily constructed or badly designed. Even so, the mechanical thresher came to be commonly used in the wheat areas of the East, and to some extent in the West. The difficulties of transportation, however, reduced the profits to the farmer, inhibited capital investment, and encouraged the continued use of treading because it cost so little.

In the late 1830s, the Pitt thresher came on the market. This efficient and inexpensive machine encouraged wider adoption of mechanical threshing, particularly in the West. By the early 1840s, most farmers used mechanical threshers and winnowers, as far west as Indiana. In 1844, J. I. Case began building his thresher in Racine, Wisconsin. He found his market in the wheat fields of Wisconsin, Illinois, and the East. The Pitt thresher began to be manufactured at Alton, Illinois, in 1847.

The first threshers simply threshed. Someone still had to separate the straw with a rake and then remove the chaff with a fanning mill or winnowing baskets. The Pitt thresher had a straw separator which shook the grain through a sieve and caught and ejected the straw. The addition of a fan for winnowing completed the development of the machine. The combined thresher–separator-winnower became important in the farm industry only in the late 1840s, however, with the expansion of the Pitt firm and the beginning of the J. I. Case company.

Farmers often owned the simple thresher. The more complicated, expensive, and heavy thresher–separator-winnower more commonly was owned by farmers who did custom work for others. Itinerant threshers took over most of the business during the 1850s in the wheat-growing areas. Horse treadmills, sweep powers, or steam engines powered the threshers. By 1860, most thresher-cleaners had devices for delivering the straw to a stack. Many of the machines also had automatic grain-bagging devices. These larger rigs became prevalent in the wheat areas, East and West, during the 1860s.

Although hand power worked the earliest machines (and these hand-operated threshers appear in large numbers in museums), such threshers had no real economic importance. Almost from the first, horse-powered threshers handled the bulk of the machine-threshed grain. The machines with their mechanisms for converting power were soon made portable. The treadmill type of horsepower conversion was most common, especially on small farms. The horse-powered sweep was used only for the larger machines.

The treadmill derived its power from the weight of the horses, and thus it could develop only a small amount of power. In contrast, the horse-powered sweep depended on the strength of the animals. In a multiple animal sweep, the master gear wheel was generally over-head so that the horses did not have to step over any shafts. But some sweeps had their master gear at the base, and the animals had to care-fully step over a shaft. These so-called horsepowers continued in use in all parts of the country long after the Civil War. Western farmers often used a huge horsepower with eight or ten horses hitched to the sweep. The time and work saved threshing wheat by using machines defies calculation when compared to the laborious, time-consuming treading.

Interestingly, all these machines—plows, harrows, planters, reap-ers, mowers, threshers—whenever invented, first came into widespread use about the same time in the 1840s and 1850s. Advances in one as-pect of husbandry required advances in other aspects. If one opera-tion became easier and quicker, it made little economic difference to the farmer unless he also could speed up other parts of production. By the 1850s, farmers had achieved a fair balance among all elements of grain production.

Technological advances not only allowed each farmer to produce more grain, but each could also increase his economically feasible area of production. This meant a total increase in the amount of grain available and decreased prices for the farmer. The adoption of tech-nological improvements helped to decrease costs. The changes re-sulted in more food, even before growers had achieved significant in-creases in yields per acre.

INVENTION OF CORN SHELLERS AND PICKERS

Of all the grains, none yields more than corn. From colonial times, corn had to be picked, husked, and shelled by hand. Seashells were often used to pull the grain off the cob; this practice might ex-plain the term "shelling." Possibly as early as the 17th century Americans had devised metal rasps and hooks for shelling; the hooks were fastened to a leather or cloth belt which fit on the hand. The hooks removed corn from the cob when scraped on the ear. To pro-tect his hand, the sheller usually wore a glove under the hand belt.

Early inventors worked out simple mechanical shellers which usu-ally consisted of drums inside a box with teeth on the drums and sometimes on the box. The teeth were usually set in a spiral like a meat grinder. As the worker turned a crank, the cob moved from one end of the box to the other, and the teeth stripped the corn off. This design was the mechanical method employed in most successful corn shellers, and several patents were given for corn-shelling devices.

Corn sheller, *ca.* 1817.

Some worked like graters, but the most successful machine used two rotating cones which forced the cob into an increasingly narrow space, thus also removing the kernels at the sloping end of the ear which were most difficult to remove. By the 1830s, various mechanical devices for shelling had appeared in the North, South, and West.

Small farm machines, including powered shellers, appeared in considerable numbers in the 1840s. In the South, planters operated shellers, commonly using a horse-sweep for power. In the North, grain dealers owned and operated the shellers. Horsepowered portable shellers were used by itinerant operators who often did custom shelling for farmers by the 1860s. As other devices for reducing the time spent in corn production came into general use, the need developed for more efficient methods of husking and shelling. A combination husker-sheller had rollers with lugs which tore off the husks and another set of rollers which shelled the corn.

Methods of harvesting corn varied; but regardless of whether the cornstalk was pulled out or cut off, the grower usually bundled the corn into shocks left in the field to ripen and dry. Especially in the South, farmers ground whole ears and cobs together as livestock feed. This practice became less common after commercial corn shellers appeared. In the late 18th and early 19th centuries, southerners often put the husked ears of corn on slotted shelves through which the corn fell. They beat the ears with long poles, and the kernels separated from the cobs. They tried to duplicate the flailing method used for wheat and other small grains, but it proved highly inefficient for corn. In the 1830s and 1840s, mechanical shellers replaced this method.

For nearly three centuries, tall shocks of corn, resembling Indian

Ingersoll's Improved Portable Hay and Cotton Press, 1857. (From *American Agriculturist*, Vol. 16, New York, July 1857)

tepees, signified the advent of autumn in America. Pumpkins were often planted between the rows and reached maturity at the same time. The Indians had originated this crop mixture and the combination was continued in many parts of America into the 20th century.

HAY HARVESTING ADVANCES

The opening of the new wheat country in the West led to a greater emphasis on animal husbandry and dairying in the East, where hay production and harvesting became increasingly important. The

hay was cut by scythe at the rate of an acre or two a day, and then it
was later raked up after it had dried out.

Farmers used wooden rakes with wooden teeth until about 1812,
when the horse-drawn rake appeared in New York. It had spread to
New England and the other Middle Atlantic states by about 1820.
Twenty years later, it had reached the West. Accounts and estimates
vary, but one man with a horse-drawn rake could probably do the
work of from six to ten men using the old hand rakes. Horse-drawn
rakes made little advance into the South where stockmen practiced
primitive ranching on uncultivated land, or they fed cattle on fodder
such as cottonseed instead of hay. Slaves provided the labor to rake
the small amount of mowed hay.

The first horse-drawn rakes were essentially large combs about 10
feet wide, with teeth 20 inches long set about 8 inches apart. The
farmer walked behind to guide his implement around or over obstructions. From time to time, he had to dump the hay out of the rake.
Later, teeth were placed on both sides of the comb, and the whole was
made to revolve Then the hay could be dumped without stopping
the horse.

The horse-drawn rakes especially appealed to farmers in the
Middle Atlantic states, particularly in New York, Pennsylvania, and
New Jersey. As subsistence farming and sheep raising gave way to
dairying and the feeder business in New England, the new rakes were
slowly adopted; but the rocky New England soil discouraged their use.

Hay and some corn were used for fodder. Cyrenus Wheeler
patented the first successful mower, as distinct from a reaper, in 1856.
His mower had two wheels and a flexible cutter bar. These mowers
could closely and rapidly cut thick stands without clogging and without tearing the sod. In the 1840s, the sulky rake with wire teeth was
introduced, but the wooden rake dominated until well after the Civil
War. The tedder, which stirred up the windrowed hay for better
drying, appeared after 1865.

IMPROVED CARE OF LIVESTOCK

All over America, livestock grazed on pastures or ranges in the summer. In the South, farmers generally pastured their animals in the
winter as well. After 1840, farmers, North and South, increasingly fed
their animals fodder through the winter. In the North they might
stable or at least shelter their stock. Animals needed less feed when
sheltered.

The early German and Dutch settlers usually had kept their
cattle in stables in the winter. The practice spread, especially as the

soil became depleted in longer settled areas. One of the major objections to allowing cattle to run wild was that their manure was wasted. Farmers in newly settled areas usually did not need the manure because of the fertility of the recently opened land. Farmers frequently expected to move West when their land deteriorated. But as manured land and better animal husbandry both increased profits, better care of animals resulted. By the 1840s, most farmers fed their cattle and many even sheltered them. This improved care took place most often in dairying.

Nevertheless, in 1852, the editor of the *Genesee Farmer* admonished farmers to improve the feeding and sheltering of their cattle. In 1853, a committee of dairymen of Herkimer County, New York, urged farmers to give cattle:

> A uniform and plentiful supply of nutricious [sic] food for them, with perfect quiet; plenty of water, requiring but little exercise to obtain a frequent supply; warm and dry stabling in winter, with quiet and careful handling; thorough and quiet milking at particular hours by the same hand; uniformity of health in all the herd of cows . . .

And so on through a lengthy catalog of good practices.

Feed in the winter season consisted of hay, corn fodder, oat straw, and sometimes cornmeal or wheat bran. For animals to be fed or "finished," farmers would generally fork the fodder into the feedlot. Even if the commercial dairymen fed their cows in a barn or stable, they moved the fodder by hand, usually from carts or sleds. Tossing the feed onto the ground was common in most of the country until the Civil War. In 1857, Orange Judd could point out: "It takes a third more fodder to winter a cow in this way, and with all the food she can eat, she comes out poor in the spring, and brings a lean calf. Either enlarge your barns or diminish your stock. Let not this barbarous practice of wintering your cattle at the stack-yard any longer disgrace the American farm."* Barbarous or not, the practice persisted for ordinary farmers through the Civil War. Large-scale changes awaited the development of improved haying equipment, which for the most part appeared later in the period.

IMPROVEMENTS IN THE DAIRY BUSINESS

The profitable dairy business made even comparatively expensive and time-consuming techniques of feeding pay. In Herkimer County, New York, the heart of the butter and cheese area, dairy farmers had

* *The American Agriculturist,* 16 (New York, 1857), 36.

altered the usual methods of animal husbandry by 1860. The editor of the *Dairy Farmer* noted in 1860: "But this increase in dairy products has not been reached alone by careful selection of stock—better shelter has been given, better food has been provided, both summer and winter, and as a result, fewer neighborhood 'bees' to lift poor, weak cows in the spring." This care was both a cause and a result of the profits of dairying.

The processing of most dairy products was done on the farm from 1783 to 1865. Processing remained overwhelmingly a farm enterprise where the first steps toward mechanization took place. Cows were milked by hand. The milk was poured into pans or tubs to allow the cream to rise. Sometimes whole milk was churned, but it was hard work and required too many churns and churnings. Milk was cooled by placing it in a cellar or springhouse. The springhouse, a small outbuilding with a stone or brick floor, was built over a spring or small stream. The water was partly dammed to maintain a low level of water on the floor where the tubs or pans were set. Until the Civil War, no significant changes were made in this collecting and cooling process. Until about the 1850s, the cream was often allowed to sour slightly before being skimmed off. If the farmer wanted the skim milk, he skimmed off the cream which was then allowed to sour. The skimming was done with a ladle before more complicated devices appeared.

Various designs of barrel or plunger churns were used by butter-makers. Commercial dairymen did not churn by hand because of the large volume of cream to be handled. By 1812 or earlier, dog- or sheep-powered churns, operating off a treadmill device, predominated in the dairy regions. The churns held about thirty gallons, and the dairy farmers usually churned twice a day. An alternative power mechanism consisted of a large wheel, some eight feet in diameter, on which the dog or sheep walked. The butter was packed in tubs of 40 pounds or firkins (casks) of 80 pounds. The dairyman might cover his butter with a cloth, but he generally placed salt on top of the butter and put on the lid. In noncommercial areas it might take as long as a month to make enough butter to fill the tub or firkin. As a result the bottom layers would be rancid, the middle strong, and the top layer reasonably sweet.

But in commercial dairying areas, farmers made fairly uniform butter, although not necessarily of excellent quality. Important commercial butter areas developed around New York City, especially in the nearby Hudson valley counties. The animal-powered churn represented the only significant technological advance before 1865, and the shift from sour cream to sweet cream butter the only methodological advance. The production of butter and its consumption reached high levels after 1812. Butter could not be transported far without refrigeration or pasteurization. Still, the surplus milk could be profitably disposed of as cheese. Cheese factories were soon built along the transportation routes.

APPLICATION OF THE FACTORY SYSTEM TO DAIRYING

In dairying, the most significant advance after the Revolution was the development of the cheese factory, followed by the butter factory, or creamery. The factory system developed slowly at first as dairymen worked out methods of operation, but then it rapidly dominated the industry. Dairying had long been a farm industry where both production and processing took place on the farm. Often groups of farmers acted as retailers. At first dairymen could not supply wholesalers because their product could not stand lengthy storage without improved technology.

The cheese factory system is supposed to have originated in Rome, New York, in 1851. The method spread to other commercial dairy regions, and by the 1870s, professional cheesemakers made most of the commercial cheese. Farmers sold milk to the factory by weight, and the factory made and marketed the cheese. The system yielded a uniform product with all the attendant advantages of large-scale buying and marketing.

Cheese manufacture changed little between 1783 and 1865. Cheesemakers put milk in tubs, heated the milk to a prescribed temperature, and put in a bit of calf stomach (rennet) to start the coagulation which resulted in cheese. The amount of rennet and the milk temperature determined the type of cheese, although other things could be added to produce variations. When the milk had curdled, the cheesemaker cut it up with curd knives and ladled it into a cheese press. The press usually consisted of a barrel, covered on the bottom with a cheesecloth. A hole in the bottom of the barrel let any remaining whey drain out. Pressure on the cheese was applied by several means. Rocks could be placed on a cover on the top of the press, but usually cheesemakers used a lever or a screw type of press. The length of time the cheese was left in the press varied between 48 and 72 hours according to the type being made. The main technological advance in cheesemaking was the use of thermometers by the 1860s. Commercial rennet was not yet used on a large scale.

After the cheese was removed from the press, it was stored in a cool place to cure. Curing sometimes took several months, during which time the cheese had to be turned over periodically. An outside covering on the cheese prevented spoilage. The covering usually consisted of old butter or lard. Cheesemakers also tried other coverings, for example, a slime of mashed vegetable matter such as beet tops.

The chief advance of the years from 1783 to 1865 was moving cheese manufacture from the farm into the factory. The factory system facilitated the application of science and new technology to the processing of dairy products. The factory could afford steam engines and machinery. Individual dairy farmers generally could neither buy factory equipment nor use it efficiently. Automation, in an early-day sense, began in the dairy industry. The factory systematized production of a uniform product. However, the factory system was not yet widespread in 1865.

PROCESSING OF OTHER FOOD PRODUCTS

A host of other farm products had to be handled and sold. Handling perishable commodities required great care and work; this raised their costs. Fresh produce could be consumed in season and, like strawberries from Florida, might be advanced or retarded in season by the distance being grown from the market. But the greater distance added to the cost. Perishability made both processing and transportation cost more. For the consumers, the greatest difficulty was getting foods out of season, regardless of cost.

Growers dried apples and green corn in the sun; they stored potatoes, turnips, carrots, and cabbages in a root cellar which was often dug into a hill or rise near the house. Squash and pumpkins kept a short while after harvest. The great appeal of the potato, aside from its high yield per acre, may well have been its excellent keeping qualities. Most food had to be stored on the farms. Few urban dwellers had cellars for root crops or cribs for corn. In urban areas even pickled meats were kept by butchers rather than consumers. Consequently farmers found the market for some commodities rapidly became glutted in season, with little marketing possible year-round. Farmers, therefore, had little incentive to grow much perishable produce unless they could spread distribution and use through the year.

The first significant development in food preservation occurred in canning in the early 19th century. In response to an appeal and a prize offered by Napoleon for a method of preserving foods for the French army, Nicolas Appert invented canning. His process involved hermetically sealing the foods in glass containers under pressure and high heat. The technique preceded the scientific discovery of microorganisms and pasteurization by about 60 years. The process, although patented, could be duplicated by anyone with glass containers, a set of lids, and simple boiling pans.

In 1819, William Underwood came from England to Boston and set up a canning plant, specializing in seafoods. He used a tarlike substance as the sealing agent on the jars. About the same time, Thomas Kensett, also from England, set up a canning firm in New York. The Appert system required the use of glass or ceramic jars, but Kensett patented a method using tin cans in 1825.

By 1839, both Kensett and Underwood had shifted to the stronger plated tin cans. These had a small hole in the top so the final plugging required only a small cap soldered into the center of the lid. Unfortunately, the food being preserved had to be cut and cooked in a size which could be pushed through the small hole in the top. Other firms soon entered the business, and the prepared foods skyrocketed in volume and in use. Emigrants and the army used canned foods most.

In 1856, Gail Borden began canning condensed milk when he perfected a method of reducing the water content of milk. Widespread use followed only after he secured a contract to supply the Union army during the Civil War.

By 1860, canners sold at least 5 million cans of commercially canned foods. By then, truck farmers sold some of their produce to the canning industry which located in the vegetable-growing areas of the East. One large canner was located in Philadelphia, and Gail Borden's plants were in Connecticut. The great impact of the industry on farming came after the Civil War, but many eastern farmers profited by the new markets even before the war. Can making was improved when Henry Evans invented a machine which stamped out tops and bottoms in 1849. Previously the tin for the sides, top, and bottom had to be cut by hand and then soldered. Later inventors worked out more efficient methods of joining the pieces.

BEGINNINGS OF REFRIGERATION

The development of refrigeration came after canning. Since ancient times men had known that they could slow spoilage by packing food in ice or in containers surrounded by ice. Because of the expense, ice packing of food was reserved for small quantities of luxury items carried rapidly from place of origin to place of consumption for the likes of Roman emperors and Spanish conquistadors.

Typical farm products were not worth the trouble and expense of ice refrigeration. Milk was barely cooled either in the springhouse of the dairy or in the home of the consumer. In the 1840s, slightly chilled milk began to be moved longer distances; and at the same time, express companies began to make small refrigerated shipments. In 1842, a caterer in Buffalo received a shipment of oysters from Albany by refrigerated express. Oyster shipments became fairly common, but oysters did not have much influence on the market for agricultural produce. The later successful express shipment of fish from Boston to Chicago showed the possibility of widening the markets for refrigerated perishables.

Entrepreneurs had thought of putting iceboxes on railroad cars and carrying game meat from the West to the East and fish from East to West in volume. Nothing came of the idea because of the difficulties of rail transport. Until all railroads had the same gauge track, delays in changing from one railroad to another made rapid ice refrigeration service on a large scale impossible. After uniform gauge had been achieved in the 1850s, refrigerated rail service became a possibility; and shippers then moved different commodities, especially meat. The meat was of game birds and animals rather than the usual farm animals.

All these methods of refrigeration depended on putting ice in iceboxes. The best solution was to develop a method of mechanical refrigeration. Several tinkerers and scientists had invented ways of making ice by compression, but the methods did not produce any significant economies. No major advance in mechanical refrigeration occurred until after the Civil War.

Food processing has long been a leader in the application of mass production machinery and methods. The assembly line originated in meat packing; bacteriological control in the dairy industry. Technological and scientific advances finally made economies of scale possible in marketing farm produce. The details varied from one commodity to another, but wheat farmers, cotton planters, dairymen, and truck farmers all profited from a combination of better processing and large-scale marketing.

CHAPTER ELEVEN

MARKETING INSTITUTIONS ❧ 1783-1865

B Y 1783, the middlemen had emerged as an important link between the producers of goods and the retailers or consumers. They became important especially at port cities where the warehousemen (middlemen) bought goods wholesale and then distributed them to retailers. Some warehousemen bought varied commodities in large volume and then redistributed them to smaller wholesalers who generally specialized in a few related products. Jobbers, who were agents without storage facilities, acted as intermediaries between manufacturers or farmers and warehousemen. Jobbers increased in number after the Revolution, although they had operated even during the war. Brokers represented would-be buyers and did their buying on commission.

As the population grew, land under cultivation increased, and urban areas became larger with a corresponding growth in the size and complexity of marketing and distribution. By 1840, the distribution of agricultural commodities involved around 3,000 firms in domestic trade and some 58,000 retail stores. Manufactured goods formed a larger portion of the commodities traded in 1840, but farm products still made up the bulk. Small home industries probably outnumbered the trade agencies in 1840, but within the next 20 years distribution facilities had come to outnumber important industrial firms. However, farmers were most numerous.

EARLY RURAL MARKETING METHODS

Between 1783 and 1861, peddlers abounded, especially in places with poor transportation. The improvement of roads and the development of a rail network enabled consumers to get to the stores, and gradually peddlers declined in importance. Restrictive laws often drove peddlers out of business, especially in cities where punitive legislation could be enforced by the rising middle-class merchants. Even so, some 16,600 peddlers still traveled the rural and urban areas of the United States in 1860.

For the farmers, the rural store was the most important distribution point. Storekeepers often became important commodity traders, buying hides, tobacco, wool, and other items; although every town had farmers' markets where farmers sold poultry, eggs, vegetables, and fruits. However, the country merchants, drawing their credit from local banks and larger urban merchants, performed various services for the farmers. These services were later taken over by the more specialized middlemen. The merchants sold manufactured goods, usually on credit. Often they acted on the side as processors of farm products, packing pork and other meats. The local miller sometimes also functioned as merchant and banker. Merchants often owned wagons, keelboats, and other vehicles. Rural merchants also served as jobbers in some areas, arranging sales and deliveries between farmers and urban processors and distributors.

From the 1830s to the 1850s, merchants performed most of the functions of middlemen and bankers as they had since colonial times. The merchants did not perform these services very well, and the interest rates they charged usually ran around 20 percent. They did not buy by grades and as a rule paid the same prices whether for excellent, average, or poor products. Consequently, the farmer had no incentive to improve the quality of his products. No wonder that Wisconsin butter arriving in New York was commonly listed as "western grease." The distant dairyman not only had no control over his product in transit, but he had no particular reason to prepare a first-class product. As urban centers grew and the number of farms rose spectacularly, the volume of business became increasingly difficult for the small-town merchant to handle. In the South the large cotton crops put especially heavy burdens on jobbers, brokers, and warehousemen.

RISE OF SPECIALIZED MIDDLEMEN
IN LIVESTOCK MARKETING

As a result of the volume of the business and the size of the markets, specialized middlemen began to appear where certain commodities moved on a large scale. The drovers are an example. The great livestock markets of the cities represented an early development of specialized facilities.

Up to the Civil War, most farmers sold their cattle, hogs, sheep, and horses to itinerant drovers. They, in turn, sold the livestock in city markets to jobbers or speculators, who then supplied a small but growing feeder industry and butchers. Another business of the drovers was selling dairy cattle to the farmers. Dairymen preferred to buy milk cows rather than raise their own heifers. Most farmers decided that it paid to buy a full-grown cow rather than carry the expense of raising a calf. Debates on the economy of raising milking stock continued after the Civil War.

Animals trailed any distance had to be fed for a while before they could be sold profitably. Jobbers and butchers both took a hand in the business, supplying the livestock to farmers and buying it back when ready for market. No important sectional differences appeared, except in the South where the planters also often operated as merchants. They also consumed livestock driven in from afar. Some historians have found evidence that handling of livestock and other commodities produced more income for some planters than their staple crops.

Until the 1860s, towns held livestock markets once or twice a week. Drovers, farmers, and butchers met there to buy and sell. Usually the buyers bought at auction. Beginning in the 1830s, the Boston newspapers began publishing accounts of the volume of business in various classes of livestock and the prices paid. Thus the papers began crop and livestock market reporting. Papers had noted grain market prices since before the Revolution. The market information columns widely disseminated information on the fluctuations of stock prices. Hog prices especially seemed to fluctuate wildy. Farmers tried to minimize the speculative aspects of this business by securing long-term contracts for hogs at specified prices. Buyers and farmers made contracts extending over several seasons. The farmer guaranteed a certain level of production and the buyer agreed to take the animals at an agreed price. These contracts became more popular during the 1830s and 1840s.

PROBLEMS OF GRAIN HANDLING AND DEVELOPMENT OF THE ELEVATOR SYSTEM

Merchants and sometimes small companies bought the farmers' grain or took it on consignment for delivery to the miller. Farmers sometimes delivered it directly to the mill. As milling improvements brought the end of the small local mill in the 1850s and 1860s, farmers increasingly used some sort of middleman, usually the storekeeper. The miller purchased the wheat and paid the merchant or his broker; then the merchant paid the farmer, unless the merchant had already taken title to the grain. In this system, the merchant rather than the miller served as a credit facility and advanced the credit the farmer needed.

From 1783 to 1850, grain was sacked and stored in warehouses which were usually owned by millers. Occasionally shippers had used barrels to transport the grain, especially wheat. Although barrels protected the wheat better, they were even clumsier to handle than sacks. This system of transfer and storage was inefficient, costly, and clumsy because the sacks or barrels had to be moved by manpower in most instances. When moved by windlass to the top of the mill or out of ship holds, the containers had to be manhandled along the way.

Grain usually was taken from ships or boats in barrels. If it had been shipped in bulk, someone had to shovel the grain into barrels which were then lifted from the hold by block and tackle. Then workers emptied the barrels into sacks which were carried to a warehouse or possibly a railroad car. Joseph Dart, a miller from Buffalo, estimated that only 2,000 bushels could be unloaded in a day, assuming calm weather and a full complement of stevedores.

Handling costs accounted for only one difficulty. The grain had to be carefully stored in warehouses to prevent spoilage from heating and other damage. To grade or inspect the grain, the sacks or barrels had to be opened and then closed. The large quantities of wheat to be marketed put almost unbearable strains upon the system. Agricultural progress and prosperity threatened to be strangled by the inefficient transport and storage machinery.

The introduction of the elevator system of grain handling prevented this strangulation. In 1785, Oliver Evans first invented and patented the elevator method, along with an almost fully automated flour mill. He used an endless chain bucket system to move the wheat from bins into a storage elevator. Then it was taken by bulk transport to a mill where another endless chain bucket system took the grain to the top of the mill. The grain was ground and emerged as flour and a waste by-product called middlings, which could be used as animal feed.

Shipping the grain in bulk to the elevators made it easier to handle. The farmer received an elevator receipt showing the amount and quality of the grain he had sold or put in the elevator. Thereafter the grain lost its identity and if the farmer wanted any back, he simply got a specified amount of a certain quality.

The new system, however, made little impression on millers or dealers until it was "rediscovered" by Joseph Dart in 1843. He improved the system and showed it to be practical. Dart's greatest contribution was adding steam power to the process. A waterpowered elevator had been tried at Rochester in 1828 and horsepowered elevators later; but these experiments had not worked well. Dart used his elevators primarily for storage rather than milling, and from the first he used steam engines to operate the chain buckets. He put the grain in bins according to specified quality, thus simplifying inspection.

The first storage elevator in Chicago appeared in 1848, and more elevators were soon built at other great terminals. This development moved milling from the rural mills to the large elevators near cities. The system spread quite rapidly just before the Civil War, but the greatest development came after the war. By 1860, Buffalo, which was the terminal of the Erie Canal, handled quantities of grain impossible under the old sack system, no matter how much that system had been expanded.

The efficiencies of the elevator system immediately affected the grain farmers. They could bypass the local monopolies of small millers and merchants. A sure market with good prices stimulated bread grain culture throughout the Old Northwest. The movement of grain onto the world market was also faster. Furthermore, the reduction in costs of transport and handling meant increasing returns for the growers, who could profitably raise grain at ever greater distances from the major markets.

The more distant farmers benefitted most by the improvements in transport and storage, for the cost reductions represented a greater percentage of their total costs. Grain growers of Pennsylvania and of Sussex, England, alike found themselves unable to compete successfully with distant producers. Wheat came to be the specialty of Minnesota; Alberta, Canada; and other remote places. Grain farmers near urban centers were forced to change their specialties.

BROKERS AND FACTORS AS MIDDLEMEN

The cotton textile industry bought cotton by grades because specific lengths and types of fiber were needed for different orders. A planter seldom produced only a particular quality, and so the product of each plantation had to be graded and separated. The textile factory required the output of many plantations. Because factories could not undertake this amount of sorting and assembling, various middlemen took over these steps by the early 19th century. Brokers handled the bulk shipment to manufacturers, and they also purchased the cotton outright or handled it on consignment. Since most of the cotton was exported to Europe, agents were needed to take care of the business.

Planters in the South and farmers in the North complained about the charges of the middlemen, but few of them could replace the middlemen. An occasional planter attempted to act as his own agent, sell to certain buyers, and carry on all of the business. Even large planters discovered that the risk involved and the time and energy needed made it far more profitable to use the brokers and factors. Indeed, the size of the operations appeared to be decisive. The largest planters found it best to consign the entire crop to someone who would try to get the best price possible. The numerous small cotton farmers in the South raised some food as well. Although nearly self-sufficient, such farmers had to buy in retail quantities at country stores. They did not gain much by going through brokers.

Wool dealers also developed as the textile industry became more centralized and the factories larger. Before 1830, many small woolen textile plants bought wool from nearby farmers or, in a few cases, owned flocks of sheep. Farmers had to act as their own merchants, carrying their wool clip considerable distances, up to 50 miles, to sell to the textile manufacturers. No regular market existed, and farmers had to peddle from factory to factory to get the best price. Beginning in the 1830s, wandering jobbers, acting as agents of a few large mills, moved through the countryside buying wool. Central markets appeared for the various textile fibers, beginning with cotton.

Before and after the Revolution, many of the cotton, tobacco, and rice brokers were in London. During the 19th century, the English brokers began to decline in prominence and American agents came to dominate the business. The Americans usually had connections with a northern banking institution, although the brokers and factors operated out of the chief southern cities. Richmond became the center for tobacco; Charleston and Savannah for cotton, and New Orleans for cotton and sugar. Other markets developed along the Mississippi River and the fall line of the Atlantic coastal plain. Vicksburg, Augusta, and Atlanta served as outlets supplying the greater markets of America and the world.

The large planters used factoring services as well as brokerage services. Factors acted as purchasing agents who bought the capital and consumer goods wanted by their customers on the best terms. These services were best provided in the larger towns where a variety of goods was available and competition tended to keep prices down. Brokers commonly took 20 percent on sales. Factors charged 2 or 3 percent for their services, but they usually saved enough for the planter to make the arrangement agreeable to both. The factors also could secure credit for the planter which he often could not obtain for himself.

This system encompassed a larger range of commodities between 1783 and 1860. The planter of the South has sometimes been criticized for the extreme institutionalism of his business and his lack of direct participation. Southerners had actually developed and used the same business methods which gained favor in the North, but more slowly. The southern planters applied these methods ahead of the northern grain or dairy industries. Bulk handling of products and sophisticated machinery for doing the business became commonplace everywhere because of the obvious efficiencies of the system.

Ultimately, for North and South alike, the source of capital was Europe. But for the southerner, the middleman between him and Europe was a northern merchant. The northerner shared a similar emotional and philosophical background with his middleman. In short, slavery and the surrounding controversy added to the southern farmers' difficulties, whether economic, political, or social. Capital in the hands of a northern merchant commonly amounted to capital in the hands of an enemy.

DEVELOPMENT OF CREDIT AND MARKETING FACILITIES

A vast country with an expanding economy needed highly developed credit facilities to carry on its business activities. The unregulated banking institutions made credit easy to get. More difficult was sorting out the bad credit risks. In 1841, the first significant credit bureau was set up in New York City by Lewis Tappan of A. Tappan and Company which became R. G. Dun and Company in 1859. The need for such an information agency had become apparent during the financial disasters of the Panic of 1837. Its exact causes are difficult to pinpoint, but wildcat banking and unsound loans certainly played a part in the panic and consequent depression. An easy money policy, generally favored in the rural areas, could succeed only with a way of separating good credit risks from bad.

The development of credit facilities, although principally affecting high finance, stock markets, and large-scale banking, also helped farmers even if they never appeared on the exclusive lists of these companies. Elimination of the largest poor risks helped protect the small country banks and thereby the credit facilities available to the ordinary farmer. Marketing and credit facilities influenced farm life and farm marketing and aided the shift to commercial farming.

BEGINNING OF COMMODITY EXCHANGES

Commodity exchanges, where commodities were bought and sold for present and future deliveries, first developed in the marketing of large-volume staple crops such as cotton and wheat. Dealers had been able to store cotton and wheat and already had grading and storing facilities when the exchanges first appeared. The exchanges placed a speculator between the farmer and the dealer and reduced the speculative nature of price fluctuations for both.

In 1848, the Board of Trade was established in Chicago for the trading of grains in a formal central market. Trading on the exchanges stabilized prices for the commodities traded. The New Orleans cotton exchange probably developed before 1837 since speculation on the exchange may have been a cause of the Panic of 1837. The exchanges helped stabilize prices for the farmer. A grain farmer could take his wheat to an elevator and receive the current price on the Chicago exchange less the cost of transportation to that market. The quotation on the Chicago Board for wheat thus became its going price, as the quotation at New Orleans for cotton became its going price. The brokers and factors of the South had already declined in importance before the Civil War, but after the war the exchanges and other marketing methods replaced the plantation middlemen.

Other exchanges for commodities, such as eggs, dairy products, and vegetables, began business. These developments were largely dependent on the ability of middlemen to handle commodities in large volumes of a consistent grade. Technological innovations in the several processing and transportation industries helped to achieve this.

CHAPTER TWELVE

LAND LAWS OF THE UNION ⚔ 1861–1914

I N THE EMERGENCE of the American farmer into the 20th century, the development of land laws after 1861 was a major element. Accounts of land laws and their implementation develop into a complicated discussion of statistics. The study of land policy has always been clogged with statistics as has most of American history. How, except in numbers, could Americans account for a country which embraced 2,969,640 square miles and 31,443,321 people by 1860?

By 1840, census takers had begun to gather statistical information on agriculture, and they filled the information schedules well enough to publish the figures in 1850. In the mid-20th century, the problem of assembling information has become so difficult that census takers now rely heavily on sampling. The statisticians merely estimate the number of horses and mules, for example. The formulation of effective policy required some statistical information, and it has been passed on to historians. In records of land distribution and tenure, historians have ample documents for study, despite a welter of archives.

Changing farm technology influenced land policy. Using 1860 technology, one farmer could handle 160 acres. Critics of western land policy, however, have observed that the 160 acres available under the Homestead Act, although enough in the humid East, would not give a farmer a living in the arid West. The issue of "enough land for a family to earn a living on" has generally centered attention on climate, water supply, and soil fertility. A later generation discovered that 160 acres was not enough regardless of climate or soil. Climate may not have been so important as the fact that a farmer had to grow an enormous amount of wheat to prosper, and technologically he was able to achieve these production levels.

Certainly the Federal Government tried to make land laws responsive to the needs of the people. The Civil War had hardly ended when Congress allowed Confederate veterans to take out homesteads in 1866. Congress also tried to achieve some measure of social equity.

A husband and a wife could not take out separate homesteads. Further-more, no one could take out a claim if he or she already owned 160 acres. This provision may have been aimed at the planter aristocracy of the South which, in spite of defeat, managed to hold onto sizeable chunks of their prewar holdings. The law also provided that each person could perfect only one claim in a lifetime. This provision could be circumvented by changing names, but Americans had so many opportunities for securing land that they did not need to resort to large-scale cheating.

In 1870, Congress provided that Union veterans could homestead 160 acres within the government sections of the railroad grants. In 1879, Congress permitted claimants to leave their claim without prej-udice for one year at a time if grasshoppers destroyed their crops. Since the worst grasshopper plagues of the time occurred between 1873 and 1877, Congress moved with remarkable speed to set this difficulty aright.

DISPOSAL OF FEDERAL LANDS

As observed before, the Homestead Act did not account for the greatest distribution of federal lands. In 1865, the public domain of the United States amounted to about 1,200 million acres. Of this, homesteading disposed of only about 147,331,000 acres. Not all of the public domain went into private hands. Of the land so disposed, the Federal Government sold 610,763,000 acres. Congress also gave large amounts of land to the states as school land grants under the Morrill Act of 1862 and a variety of other laws.

The states almost universally wanted title to all the land within their boundaries. They sometimes came close through a variety of land grants for education, transportation, river improvements, swamp-land, and others. Arkansas, Louisiana, Michigan, and Minnesota man-aged to get title to one-third of the land within their boundaries, and Florida got two-thirds of her land. This land available to farmers no longer counted as part of the public domain when released to the states. Eleven public domain states took some 1,769,000 acres, and 27 or so states took 8 million scrip-acres. This came to less than 1 percent of the land homesteaded or sold by the Federal Government after 1862. The college land scrip and the other types of state land fell mostly into the hands of speculators, instead of going directly to farm-ers. The exact amount that went to speculators has not been deter-mined. A great deal of work remains to be done to find all the in-formation and analyze it.

In 1870, under the Preemption Act, Congress allowed the use of college scrip to pay for land. Under this act which remained on the books until 1891 the government disposed of a sizeable amount of land and assisted the farmers. But exact accounting is difficult because this land might be considered simultaneously as land preempted and as land given to the states. The states which had scrip sold it to raise revenue to support their land-grant colleges. The scrip brought varying prices at different times and places. States realized from 50 to 90 cents an acre. Although the states got a large amount of land, in terms of the total public domain, the states received a small amount.

GRANTS TO RAILROADS

Railroads in the West received part of the federal land as a subsidy. The facts are confusing and have resulted in some interesting calculations on the actual amount of land given away. However, most historians have used their calculations chiefly to condemn the railroads.

Two items help confuse the story. First, the railroads had to return land when they failed to construct, and large acreages did revert to the Federal Government. Second, the states gave large acreages to the railroads, and these were now and then counted both as gifts to the states and gifts to the railroads. By 1865, some 102,272,573 acres had been granted to the railroads, starting with the Illinois Central. Not all of these acres remained in railroad hands. By 1877, the railroads had received an additional 53,232,000 acres from all sources. By the end of World War I, the railroads had received around 129 million acres of land from all sources. This was not a very large proportion of the amount sold by the Federal Government.

Prior to the Homestead Act of 1862, the Federal Government had doubled the price of land in its sections of the checkerboard grants. The government thus assured the railroads of the sale of their land. The law also assured the Federal Government of the unearned increment in the value of the public lands within the railroad grants. After Congress passed the Homestead Act, the government held its sections off the free list until the railroad land had been taken up. Then for a number of years, farmers could homestead only 80 acres instead of 160 acres. Still, the land could always be purchased at a minimum price of $2.50 an acre. The federal lands rarely went for so little as the minimum price.

Congress next extended the belts of land which could not be homesteaded until the railroad had sold its land. The law reserved land from 60 to 120 miles on either side of each right-of-way. This nuisance law ended in Cleveland's administration in 1887. The reserved federal lands were taken off the homesteading list even for projected lines never built. The granting of land to railroads effectively ended in 1871, and the last western railroad, the Great Northern, went forward without any federal land. It did, however, get help from the states and from Canada.

For most settlers a homestead claim served to get them started in farming. Homesteading also enabled an immigrant to get the lay of the land before investing too much. Free land or not, people with little or no experience in farming often made serious mistakes and failed as farmers. The free land enticed some into areas unsuitable for farming or for certain types of farming.

Yet in the eastern parts of western states—the area generally known as the prairies, as distinguished from the plains—the act worked fairly well. Until the 1880s, settlers generally found the amount of land available under the act sufficient for farming in the more humid areas. The commercial farmer had to locate near routes of transport— usually a railroad. The advantage of being close to transportation more than compensated for the higher price of land.

A farmer who had so little capital that he could not buy a farm usually could not succeed with a free farm, for homesteaders led the parade of bankrupt farmers. The railroads had no interest in creating zones of poverty along their rights-of-way or in not disposing of their land. They wanted customers with money. The railroads could sell their land for fairly high prices. Reliable average figures are not readily available, but the railroads apparently disposed of the bulk of their land for about $4 an acre.

In the early days, railroads often sold the land for even lower prices and sometimes practically gave the land away to the first settlers. Some roads paid their workers partly with railroad scrip. Some Irish and Swedish immigrants located near the railroads because of this policy. In other cases, the sale of the scrip helped pay railroad construction costs. By 1865 at the latest, the railroads had already developed a generous credit policy, usually calling for one-tenth down and the rest in ten or eleven years at a relatively low interest from 6 to 7 percent. The railroads could not sell their land if they did not sell on credit. On the other hand, when grasshoppers or drought ruined settlers, the railroads could foreclose and advertise already broken land for sale.

Railroads provided free transportation to their lands for settlers, often gave them temporary shelter in tents or shantytowns, and generally offered services to assist prospective customers. The railroad men welcomed colonies of settlers, often led by their ministers, who came from the Netherlands, Sweden, and other places. In 1874, the Sante Fe sold 60,000 acres to German-Russian Mennonites in Kansas. Most Mennonites came in colonies as did various other national pietistic groups. In all cases, railroads treated the sale of the land as incidental to establishing longtime customers.

Many railroads and state governments kept agents in Europe and in American ports of entry to attract immigrants to railroad or state land. These efforts succeeded remarkably well, especially when combined with the generous credit offered by the railroads. Hordes of immigrants from Germany, Scandinavia, Russia, and Canada came to the prairies. At the same time, internal migration took place. Native Americans actually made up the bulk of the new settlers: yeomen farmers from the South and Yankees from the Midwest, the East, and New England. They usually moved along parallels and into lands of similar climate.

The railroads not only encouraged and facilitated this movement of farmers, but they also carried on agricultural experiments and spread information among their settlers. The Northern Pacific, for example, led the way in the development of "bonanza" farming in the Red River valley of the North. This development (about 1876 to 1892) not only generated traffic for the roads, but also illustrated the agricultural potentiality of the land. The bonanza farms eventually gave way to small-scale enterprises, as the depression of 1893 broke the corporate farms and the smaller operators moved in and took over. The land had by then passed from the Federal Government to the railroads and from bonanza to small farmers. Private transfers clearly surpassed the initial means of disposal in importance. The actual settlement might have been considerably delayed, and less favorably for farmers, had the railroads not played their part and the bonanza farmers theirs.

DEVELOPMENT OF DRY FARMING

Beginning in the 1880s, James J. Hill of the Northern Pacific Railroad urged the development of dry farming in the semiarid West. Largely through railroad efforts, a Bureau of Dry Land Agriculture was set up in the Department of Agriculture in 1906.

Many scientific advances occurred between 1865 and 1914, but among the most impressive was that complex of discoveries lumped under the rubric of dry farming. Agrarian enthusiasm for dry farming, combined with a congressional hope for getting the rest of the public land into private hands, led Congress to enact the Enlarged Homestead Act on February 19, 1909.

The main idea of the dry farming complex was allowing half the land to lie fallow each year to accumulate moisture. The farmer thus grew one year's crop on two year's water. This system was probably discovered in antiquity and used for securing moisture rather than for restoring fertility. But by the settlement of America, the ancient reason for rotations and fallowing had been forgotten. Americans in dry lands rediscovered the chief reason for fallowing.

The Homestead Act had allotted 160 acres to a farmer who met its requirements. But if a farmer could cultivate only one-half his land in any given year, 320 acres were needed to homestead. The Enlarged Homestead Act of 1909 provided that the settler only had to cultivate 80 acres successfully for five years in order to secure title. Congress specified the nature and location of the land so no one could secure 320 acres in a fertile, humid area.

The land available under the act included "nonmineral, nonirrigable, unreserved, and unappropriated surveyed public lands which do not contain merchantable timber . . ." in the states of Colorado, Montana, Nevada, Oregon, Utah, Washington, Wyoming, and Arizona. The Public Land Commission included some 30 million acres in this category, but not much of it was tillable. Between 1909 and 1910, homesteaders entered over 18,329,000 acres, but only a small number of these entries were ever perfected. In 1912, Congress reduced the time of residence from five to three years. The exact amount of land that homesteaders took out and perfected under the act by 1914 is difficult to estimate—perhaps 3 million acres.

Dry farming never achieved what it was supposed to: the desert never "blossomed like a rose." But dry farming, in combination with other elements, did contribute to survival and eventually to prosperity. In terms of commercial agriculture, farmers fared best on railroad land. Buying from a railroad was one of the best ways of securing a farm in many parts of the country.

With time, technological advances, and economic changes, farmers on the prairies and the plains eventually prospered. During the first stage of settlement, however, life was often difficult. Farmers and politicians proposed various corrective measures. Protest movements were led by the Grangers, Greenbackers, and Populists, who sought solutions. As it happened, more effective measures unexpectedly rose from the ranks of the Republicans and Democrats. The older parties tried various expedients to meet the climatic peculiarities of the West.

Congress apparently believed that settlers preferred land with trees on it. Many people also believed that trees would improve the climate of the drier parts of the prairies and plains. On March 3, 1873, Congress attempted to foster tree growing on the plains by passing "An Act to Encourage the Growth of Timber on the Western Prairies." Under the act any person who planted 40 acres of trees on the claim and kept them there for ten years could receive 160 acres. Some farmers took land under the act, with varying success. In 1878, Congress amended the law to reduce the acres of trees to 10 acres, and the number of trees from 2,700 at the start to 675 at the end. No federal official was likely to count the trees.

Farmers who used the act generally planted their ten acres of trees as shelterbelts or windbreaks The later development of the shelterbelt program probably developed out of the successful experiments of farmers under the Timber Culture Act. The act did not, however, stimulate a rush to the plains, and so Congress thought it had failed. Only 9,000 entries had been proven as of 1890, and Congress repealed the Timber Culture Act in 1891. But even after repeal, farmers who had taken claims could continue to prove their claims, and they had proven over 65,000 entries by 1904. Altogether farmers took over 10,900,000 acres of land under the Timber Culture Act of 1873. The farmers who took advantage had to have capital for planting and maintaining the trees; but those who could use the act benefitted not only in the land acquired, but in the resulting shelterbelts.

Meanwhile, as farmers moved onto the plains, conflicts arose between cattlemen of the open range and the settlers, or "nesters." The conflict has been immortalized in fiction, motion pictures, and television shows. These sometimes dreadful fights concerned the predominantly eastern Congress, which adopted the idea that it could stop or reduce the conflicts by quickly flooding the plains with nesters. These farmers would deprive the seminomadic herders of their free grass. Since the Homestead Act did not offer enough land for successful farming on the plains at the existing technological stage, Congress altered the land laws.

DESERT LAND ACT

Under pressure from President Ulysses Grant and his secretary of the interior, Congress passed the Desert Land Act on March 3, 1877. The act applied especially to California, Washington, Oregon, Idaho, Montana, Wyoming, Dakota, New Mexico, Arizona, Nevada, and Utah. Under the act the settler could buy 640 acres at a price of $.25 per acre, if he would irrigate the land within three years after filing his claim. The applicant had to be a citizen, could not assign his claim, and could make only one claim in a lifetime.

To a considerable extent the Mormons had used the Homestead Act and its commutation clause to secure land with legal title for the Saints. At the same time they maintained the small farm system which they had begun in 1846. Church leaders appointed those who would take out claims and then divided the land among the actual settlers according to the plan of the commonwealth. However, the Desert Land Act offered them little help.

Other settlers usually lacked the capital to build the necessary irrigation systems. The land cost too much when the expense of the irrigation works was added. Some colonies were started, and water companies were formed to supply irrigation water to settlers, but these proved almost universally unsuccessful. In general, the farmers refused to pay the companies but took their water. Such tactics eventually brought the water companies to bankruptcy. The farmers usually ended up in possession of the water company and reformed it as a cooperative. All of this worked well enough in providing irrigation water in the few places where men made the attempt, but the eventual outcome discouraged outsiders from investing capital and the farmers usually did not have the capital themselves. Within a few years, Congress found that the Desert Land Act had not worked very well.

The act proved beneficial to a handful of speculators. The cattle barons could secure title to land around streams. The big cattlemen thus effectively cut out lesser stockmen. They secured a certain water supply, prevented irrigation of the remaining land, and grew some supplemental feeds on the acres which they farmed under the act's requirements. The act accomplished exactly the opposite of what Congress intended by aiding the cattlemen instead of the sodbusters.

As early as 1885, Land Commissioner Andrew Jackson Sparks recommended repeal of the act. He was fired instead. Congress, however, continued to debate the matter. Finally on March 3, 1891, President Benjamin Harrison signed an act which restricted the area available under the Desert Land Act to 320 acres. This still failed to attract settlers to the plains, but it offered enough incentive for most cattlemen who could have their cowboys take out claims if necessary.

In 1904 and 1905, the Public Lands Commission reported that those who took land had almost never irrigated as the law required. Of 37,000 claims filed by 1900, only 11,000 had been proven. The commission also discovered that the chief beneficiaries were cattlemen and land companies. The act allowed corporations to obtain land under the act, and many of these corporations with identical stockholders obtained large tracts of land. Few individual farmers ever benefitted from the act. The law apparently remained in effect until the passage of the Taylor Grazing Act in 1934. The Desert Land Act did not result in a massive seizure of the public domain—only around 10 million acres were ever secured under it.

Changes in agricultural technology and science which took place between 1865 and 1914 had some immediate impact on land legislation. The technological changes increased the productivity per manhour and allowed the farmer to handle increasing acreages with an actual reduction in work force. Scientific advances brought about the possibility of larger yields per acre, but this potentiality has been obscured by the statistics. If an acre of land yielded little or nothing and someone found a way to cause the land to yield regularly, an increased production per acre resulted. Statistically the amount still seemed unimpressive because the total yield was only average or below average. Was an average wheat yield of 15 bushels to an acre scientific progress? It was, if one understood that the yield had been zero before.

ESTABLISHING NATIONAL FORESTS
AND THE FOREST SERVICE

Not all of the land available to farmers actually went to them. The New Englanders of the 17th and 18th centuries had provided the common. Long after, they continued to set aside the village green to be used as a park and for military drill. The idea of a village common and green spread across the Midwest and even to the South. The Mormons carried the system to Utah. States sometimes applied the principle in setting aside state parks and forests, especially in the 1870s and 1880s. When the Federal Government finally accepted the idea, Congress implemented the system on a vast scale.

New York led in establishing state forests before the Division of Forestry was created in the Department of Agriculture in 1886. Europeans had had such forest reserves since at least the 16th century. America, however, had such abundant forests that few people had given serious thought to the depletion of this resource. The earliest efforts to save or replant forestlands came in the more densely populated states. The forest conservationists had greater emotional than economic needs when they first achieved an effective response to their demands.

The Federal Government first put the Division of Forestry under the control of Bernhard E. Fernow, a forester trained in Europe. He almost single-handedly aroused public and congressional interest in forest protection, which resulted in the establishment of national forests in 1891. The act of 1891 reserved some of the public domain for forestry purposes, under the supervision of the Department of the Interior. In 1897, Congress empowered the Bureau of Forestry to protect the national forests from abuse and illegal use. The creation of the forest reserves and their effective administration took place only during the presidency of Theodore Roosevelt. Under President Benjamin Harrison, some 13,000 acres, mostly timberland, had been withdrawn.

The most ancient meaning of the word "forest" was reservation. The king's forest in France or England was land set aside for the use of the Crown, usually for hunting. As trees became scarcer, forested Crown land was set aside. When Americans used the word, it had come to mean almost the same thing as timberland. By allowing the government to keep certain lands from being used either for farming or wood gathering, precedent had been established for government withdrawal of any land from private use, with or without trees.

President Theodore Roosevelt, using his authority to withdraw public lands from settlement, withdrew large amounts of grazing land. Effective regulation remained a problem—the two most corrupt services in the Federal Government were allegedly the General Land Office and the Office of Indian Affairs. Both were located in the Department of the Interior. In 1905, Congress transferred the administration of the forests to the Department of Agriculture, which had long had a Division of Forestry. This was renamed the Forest Service. In the same year, Congress allowed foresters to arrest those who violated the laws and regulations for the forests This meant the arrest of herdsmen who illegally grazed the reserved land.

As control of forestry affairs tightened, the government increased the amount of land so regulated. From 1905 to 1906, the forest reserves rose from 85,694,000 acres to 107 million acres. The tightening of control, enforcement of the law, and the additional withdrawals displeased many western congressmen. They held that the withdrawals hindered the business of their herding constituents and threatened to inhibit settlement and agricultural advance. Some congressmen had personal interests in the economic exploitation of the resources and so had an additional reason to object to the course of events under Roosevelt. In 1907, Congress ended the authority of the president to withdraw land for forests in six western states. Roosevelt could not hope to veto the act successfully, and so he signed the act. But before he did, he created 21 new forests in the proscribed area and increased forestland in the region by 16 million acres. Altogether, between 1906 and 1907, the total forest area had been increased by 44 million acres.

Roosevelt left office in 1909, and the next year Congress passed the Withdrawal Act. Congress again gave the president authority to withdraw forestlands belonging to the public. Congress still refused to make leasing of the forestlands legal. Thus nonuse rather than wise use became the forest conservation policy imposed by Congress. Half a century elapsed before Congress approved regular harvesting of forests.

Changes occurred sooner on the grazing areas of the forests. Especially after 1910, forestlands were reclassified, or released to entry. This mollified congressmen and westerners alike. At the same time, congressional refusal to permit leasing or long-term grazing rights acted as an impediment to increasing the forests. In 1912, Congress specified that the president could not thereafter call nontimberlands forests. The law also ordered the president to speedily return treeless "national forests" to the public domain. There was no way to force the executive to implement the act, and the president did nothing. Had the act been carried out, about half the U.S. forest would have been returned to the public domain.

As early as 1906, the Forest Service charged a grazing fee for cattlemen or sheepmen using the grass of the forests. Foresters regulated grazing, including the number that stockmen could run and who could run them. The foresters reduced overgrazing, but the low fees charged gave those who used the forests an advantage over those who owned or leased private range. On the other hand, the forest users had to pay something into the treasury which users of the rest of the public domain did not. Cattlemen and sheepmen protested the fees charged by the Forest Service. Those without grazing permits especially complained because the fees came to less than those charged by private landowners. The Forest Service tried unsuccessfully to raise fees before 1914. But the Forest Service policies impeded the movement of farmers and at the same time gave certain herders considerable advantages in grazing rights. The herdsman in a national forest had some assurance of tenure (but not ironclad) so he was better off than the livestock producer operating mostly on open range. In retrospect, farmers never lost any appreciable amount of excellent farmland to national forests.

By the time the forest reservation system was put into effect, the American farmer was often overproducing. The removal of additional farmland may have helped the nation and its farmers. The vast frontier expansion had ended, although much land still remained for exploitation.

The forest conservation movement began in the East. The record shows clearly that most westerners opposed conservation or the Forest Service, or both. Westerners consistently wanted less federal control, not more, well into the 20th century. Reformers were usually viewed as perpetrators of the evils of conservation: federal regulation and control of the public lands. Conservation of forests (with or without trees) found most support in the urban East. But easterners had no public lands left to conserve.

On March 1, 1911, the president signed the Weeks Act which provided federal money for the purchase of eastern forestlands for national forests Under the amended and extended act, the forests of the eastern states slowly grew in size and importance. States sometimes pursued similar policies, using state money. Sometimes abandoned farms were included in state or national forests. In the East, the forests served primarily as recreation and wilderness areas. The eastern forests were crowded with tourists, vacationers, and weekenders. Whatever westerners thought of federal reserves, the eastern urbanites appreciated their forests.

The total land acquired by farmers, cattlemen, and other agricultural users cannot be calculated. The inadequate records and the vast number of transactions allow only estimates. Probably from 150 to 300 million acres of federal land were transferred to farmers between 1865 and 1914. These acreages came from the federal public domain, but other land was sometimes figured in the disposal accounts. For example, Texas kept its public domain, and Texas farmland or grazing land did not enter into the federal statistics.

USE OF SCHOOL LANDS

From 1785 to 1847, each state received Section 16 in every township for the support of education. From 1848 to 1890, the new states were given sections 16 and 36, and later states (from 1894 on) received sections 2, 16, 32, and 36 or the equivalent for supporting schools. Congress expected the states to sell the land and use the profits for the support of education. But the states sometimes ignored or defied Congress and leased their school lands. New York, for example, secured lands in Wisconsin under the Morrill Act of 1862. Cornell University kept the timberlands and managed them very profitably. This was clearly different from what Congress intended. By the 1890s, however, Congress merely limited to 640 acres the amount of state school land any one person could lease, as in the case of Wyoming. Congress put no leasing limits on other state lands. Mineral (particularly oil) leases on state lands added to the money available for education in some western states.

Most states sold their land at various prices and quantities to secure revenue. The states had absolute freedom in disposing of the land, and many private fortunes were made and lost speculating in these lands. However, the states seem to have realized little from the sale of their lands. In 1875, Congress set a minimum price for which the states might sell their lands Thereafter, all states except Utah had such minimums established. Such minimum prices removed state school lands from competition with federal and railroad lands. The prices allowed by Congress usually ran between $3 and $5 an acre. But Congress allowed terms of as much as 30 years and down payments of as little as 5 percent. Interest on the balance could be as low as 4 percent. In the disposal, the acts favored crop farmers because the scattered lands were not easily managed or consolidated for grazing.

In general, the land legislation after 1865 facilitated and possibly accelerated settlement of the prairies, and after about 1870, land laws tended to encourage settlement on the Great Plains. The Plains was chiefly cattle country and acted as an impediment to settlers crossing to the Pacific Coast. Congress passed a series of land laws intended to make the settlement of the Plains more attractive. The laws of the 1870s must have had some effect. The population on the Plains increased: 1870, 1,481,603; 1880, 3,549,264; 1890, 6,053,545.

The terrible drought years between 1887 and 1897 caused a retreat from the Plains. The population growth for the decade 1890 to 1900 reflected these conditions when the number of people in the Plains states (chiefly in the more humid eastern areas) rose only from

6,053,545 to 7,775,430. This represented a sharp decline in the rate of settlement.

Spurred by land legislation and improved dry farming techniques, a second advance onto the Great Plains took place between 1900 and 1910. The flood subsided after 1910, and new immigrants mostly replaced those forced off the Plains. A trickle continued, especially in the northern Plains through 1912. It had become apparent that land policy alone could not solve all the problems of farmers on the Great Plains. Technology, science, and marketing changes also had to be adapted to their needs.

CHAPTER THIRTEEN

SELLING THE ABUNDANCE ⚞ 1861–1914

T ECHNOLOGY and science seemed to dominate American agriculture from 1861 to 1914. Farmers had to sell their produce at prices high enough to afford the new technology. Therefore, profitable and effective marketing was necessary for the adoption of new methods, equipment, chemicals, plants, and animals.

Americans could usually sell their agricultural surplus on the world market. The greatest profits came during difficult European periods, such as the Franco-Prussian War. As the 19th century ended, the American farmers' need of the European market gradually declined. From the 1890s to the eve of World War I, they came to depend on American urbanites. The domestic market took an ever greater proportion of American agricultural products. After a period of economic distress from 1867 to 1898, farmers experienced a rare time of real prosperity. A later generation called the period from 1898 to 1914 the "Golden Age" of American agriculture.

In 1860 (before Lincoln even took office), southern banks began withdrawing their deposits from northern banks. Southern planters and others repudiated their debts to northern merchants. The northerners lost at least $300 million before hostilities had even begun. The complete loss of southern accounts with the secession of the South increased northern financial and trade distress through 1861. The withdrawals strained the cash reserves of many banks and commercial houses, and the New York banks had to stop specie payments by December 1861. The Federal Government followed with similar action. Commerce did not cease but it stagnated in 1861.

Although the depression hit hard it did not last long. The depression caused the collapse of perhaps 6,000 commercial houses and caused more havoc than the depression of 1857. In the heavily agricultural West, where credit and market facilities were especially necessary, banks closed in astounding numbers. Merchants and some manufacturers responded by providing short-term credit. It served

151

well enough in most of the northern farming communities until new banking regulations and new fiscal and monetary policies rejuvenated the economy. The issuance of greenbacks by the Federal Government made money available; and although currency inflation increased as the war went on, the farmers at least could pay old debts. Northern farmers also could borrow more money to increase land holdings and to purchase the new equipment available. Country banks and other financial institutions that survived the depression began extending credit to farmers. At the same time, inflation and unceasing war demands produced prosperity in most northern businesses. As in other times of inflation, the farmers benefitted first although they did not always profit most.

Rising profits, combined with labor shortages, produced marked changes in farm technology. In turn, profits rose more. Army enlistments drained off farmers of the North and South. New job opportunities in mines and factories also depleted the farm labor force. Farmers had to find more efficient methods or use more machinery, and they could afford the new machines under the prevailing economic and market conditions. As a result, production rose especially in two areas: commodities to which the new technology could be best applied and wartime commodities produced in the North.

In 1862, wheat production reached a new high of 177 million bushels. This level of production largely reflected the application of previously available machines and implements, now suddenly used on a large scale. The market called for more wheat, particularly for international trade; but market demands did not explain all of the increase. In comparison, corn production fell slightly despite the need for it in livestock feeding.

Livestock production rose in response to the demands of armies in the field. The numbers of hogs and cattle (for meat and leather) increased slowly. Sheep became much more profitable. The number of sheep on northern farms doubled and wool production trebled as northerners sought to produce a substitute for southern cotton. California and other places did not provide any appreciable amount of cotton during the war. Flax could not be used on cotton textile machinery, and so that crop could not serve as a substitute.

Sugar presented special problems since the South was its principal source. Even heavy importations could not fill the demand. In Wisconsin and other northern states, the government tried to stimulate the production of sorghum as a substitute for cane sugar with limited success. Although sugar remained scarce throughout the war, its shortage was a minor problem because not much sugar was used then.

Other changes also stimulated increased northern food production. The northern rail trunk lines were completed. European crop shortages in 1860, 1861, and 1862 created new markets for food in addition to army demands. The northern population grew during the war from natural increase and from immigration. Some 801,723 new consumers immigrated between 1861 and 1865 out of a total northern population of 24 million in 1865. This was also the time of greatly increased urbanization.

INFLATION AND THE LEGAL-TENDER ACT OF 1862

Probably the most significant development in the economic revival after the depression of 1861 was the creation of a currency which met the needs of the farm community. The inflation apparently hurt nearly everyone except industrialists and farmers. The government bore higher costs for the war. The professional classes suffered as creditors and as businessmen who could not push fees up fast enough to cover inflation. Workingmen found that their wages did not rise fast enough to keep up.

The inflation began with the Legal-Tender Act of 1862. The end of specie payments by banks in 1861 had created a demand for paper money, especially by farmers. On February 25, 1862, Congress provided for the issuance of $150 million worth of legal-tender notes, paper money which had no backing except the authority of law and the confidence of the people in the government. Two later acts increased the total amount issued to $450 million. To a population unused to fiat money, the unbacked paper currency seemed unsound. It rapidly depreciated in value, although a general price inflation took place at the same time. By 1864, 100 paper greenbacks equalled only 39 gold dollars.

The officials usually explained the legal-tender money would allow the government to meet the costs of the war. For the first time, the government sold bonds to the general public on a large scale, although it acknowledged the obvious danger of inflation and its damage to creditors. But the government needed the money, and most farmers wanted it, too.

At the same time the Federal Government added a vast array of excise taxes. Salmon Chase, U.S. Secretary of the Treasury, followed the principle of taxing every commodity that could be taxed. This raised prices, speeded inflation, and generally increased the war costs. Gold and silver disappeared from circulation.

Agricultural commodities led in the price rise. Farmers shared in the profits of inflation; but to meet their increased costs, they usually borrowed and hoped to repay with less valuable money. In the meantime, they added to their landholdings and to their capital investment. The index of prices and wages shows what happened in general.

Year	Prices	Money wages
1860	100.0	100.0
1861	100.6	100.8
1862	117.8	102.9
1863	148.6	110.5
1864	190.5	125.6
1865	216.8	143.1

EFFECTS OF THE CIVIL WAR ON THE SOUTHERN FARMERS

From the outset of the war, the small farmers of the South (comprising well over three-fourths of the total southern population) did not stand to gain much from an independent Confederacy. Slavery only made sense to those who already had slaves or had some prospect of getting them. Simply ending the protective tariff (especially that of 1857, which presumably burdened southern farmers and made them rely on northern manufacturers for capital goods) would not create the necessary conditions for industry. However, the view that the South had little or no industry was probably truer of heavy industry. It manufactured agricultural implements and machinery, so the tariff was not an issue for farmers, insofar as the means of production were concerned.

States rights versus the power of the Federal Government must have seemed an empty issue to farmers who often lacked routes of transportation, and for whom federal aid might well have seemed the only easily available solution. Setting up their own nation hardly made sense to farmers on the whole. The causes of the Civil War were emotional and possibly philosophical, but not based on practical economics.

No matter what southern farmers thought of the military spirit of the supposedly "soft" northerners, the Confederates knew that the North had the superior navy. Ultimate relief from the damage wrought by northern naval superiority was supposed to come through diplomacy and from general economic and trade policies. Southern farmers depended almost totally on the European metropolitan market. They expected no changes in agricultural policy that would alter their dependence. As it turned out, neither southern diplomacy nor trade policy helped. The southern farmer would have emerged from the war in miserable shape even if the Confederates had won.

From the beginning the northern blockade of the South hurt

southern farmers. Between 1860 and 1861, cotton exports fell about 17 percent; and during the next four years, cotton exports came to around 13 percent of the amount exported in 1860. Blockade-runners managed to get through the northern cordon, especially from the ports of New Orleans, Mobile, Charleston, and Wilmington. But Union forces captured the first two ports in 1862, which reduced Confederate outlets. Some cotton went overland to Mexico and thence out by way of Mexican ports, chiefly Vera Cruz. But, southern producers could get little out this way, and what they did export brought little profit. After the Union got control of the Mississippi and thus split the Confederacy, only cotton from the western side could be moved out by way of Mexico. The farmer and planter made very little on their crops even if they did reach a foreign market; the amount of cotton and tobacco sold abroad in 1863 came to only about $3 million in gold.

Actually, the South could have shipped out more cotton during 1861 and 1862 because the blockade was not fully effective during the first year or so. But the Confederate government forbade the export of cotton, hoping to produce a cotton famine in England and France which would force these nations to recognize the Confederacy and to lend it military assistance. The famine did not develop soon enough because of the unusually large crop of 1860, much of which the English and French still had on hand in 1861. The famine did develop by 1863, and cotton rose to 82½ cents a pound in England. By that time, however, the North had an effective blockade and had begun to win some battles. Europeans no longer looked for a Confederate victory, so the Confederates never achieved their diplomatic objective, and they could not take advantage of the high prices of 1863 to 1865.

Cotton growers had to store their crop in hope of a southern victory. The Confederate and state governments took some of the cotton for taxes and stored it against eventual victory. All of this cotton eventually fell into northern hands.

Cotton continued to be grown, especially on larger plantations, but the small farmers turned to other crops. The cotton crops of 1862 through 1865 combined came to less than the crop of 1861 alone. However, the shift away from cotton, although substantial, never became as great as the Confederacy needed. Farmers in a position where their food crops might be confiscated or paid for in Confederate money preferred to raise something that the army could not eat. Planters feared slave uprisings and so grew cotton in order to keep their slaves busy. Furthermore, a shift to another crop would have required retraining of the slaves.

On the whole, the southern farmer produced enough food for himself, for the armies, and for the urban populations. But the transportation system in the South was inadequate in many places and the

Old style screw press for baling cotton, *ca.* 1850.

problem of payment arose. The money of the South depreciated even more rapidly than that of the North, mostly because the South started with less and inflated more to meet the war costs. The southern farmer had little opportunity to spend the legal tender on equipment and land. Little land could be acquired in the beleaguered Confederacy. Implements and machines were soon unavailable. The small amount of iron and steel had to be used for military purposes.

The small (but not inconsiderable) industry of the South had to be turned to making weapons and other war materiel. The southern farmer could not replace his farm equipment; seldom could he repair it. The farmer could fix wooden parts readily enough, and the local blacksmith or the farmer could find enough metal to make small repairs. However, replacing an entire plowshare, for example, was usually impossible. In contrast to his northern brother, the southern farmer had no opportunity to benefit from inflation. The Confederate farmer had nothing he could buy with his potentially higher income. Difficult transport problems made marketing a surplus nearly impossible and prohibitively expensive.

TABLE 13.1.   Exports, 1860–65

Year	Wheat & Corn (North)	Beef, Pork, & Their Products (North)	Cotton (South)	Tobacco (South)
	bu		lb	
1860	21,462,000	161,211,000	1,767,686,000	173,844,000
1861	64,348,000	184,829,000	307,516,000	168,000,000
1862	81,619,000	395,585,000	5,065,000	116,723,000
1863	75,262,000	532,203,000	11,385,000	118,750,000
1864	46,615,000	362,461,000	11,994,000	113,384,000
1865	26,761,000	190,334,000	8,894,000	161,355,000

SOURCE: Nourse, *American Agriculture and the European Market.*

The southern farmer also faced the loss of his crop and every-thing else through capture by northern armies. The Confederate army successfully invaded the North from time to time, but the pres-sure from the South was not long lasting and the war increasingly shifted to the South. Cattle, hogs, and other livestock disappeared as the Union armies passed. Some southern states lost one-third to one-half of their animals. Implements were in disrepair, and because of a shortage of lead, most barns and houses had gone unpainted at least four years by 1865. The southern farmer was in extreme distress as the war ended; all that he had was his land, and he could not be too sure of that. The slaves were freed, the money declared worthless, and the farmer had neither cash nor credit when Robert E. Lee finally surrendered.

During the Civil War, exports of cotton, rice, tobacco, and sugar had practically ceased. This damaged the southern economy and hurt the North as well. Northern intermediaries in the export of cotton, who had depended on the crop to obtain European capital, recovered some by expanding their other agricultural exports. Table 13.1 shows what happened to exports during the war. After the war, southern crop exports rose rapidly, although northern moneylenders took most of the returns. Northern agricultural exports also shot up, stimulated by the expanding industrialization and urbanization of Europe and the development of rapid and inexpensive transportation. Improved handling and transfer of products also speeded the process.

POST–CIVIL WAR EUROPEAN MARKETS

A veritable flood of cheap American food had entered Europe by 1875, and the British and other European farmers felt the competition. The metropolitan market of Europe grew after 1860 as the population of

TABLE 13.2. Exports, 1867–1901

Years	Wheat	Corn	Beef & Its Products	Pork & Its Products	Cotton (500 lb = 1 bale)
	bu			*lb*	
1867–1871	35,032,000	9,924,000	54,532,000	128,249,000	902,410,000
1897–1901	197,427,000	192,531,000	637,268,000	1,528,139,000	3,447,910,000

SOURCE: Nourse, *American Agriculture and the European Market.*

155 European cities rose from 19,891,183 in 1860–1865 to 51,236,301 in 1910–1915. The most important increases were in 39 cities of Great Britain, 12 of Italy, and 13 of Russia. The increase in the size of the cities of Europe accounted for much of the prosperity of the American farmer.

By five-year averages, the figures for exports show some astounding developments betwen 1867–1871 and 1897–1901—the period between the end of the Civil War and the beginning of the "Golden Age" of American agriculture.

In the 19th century, European agricultural production rose too slowly to supply the growing market. The farmers and ranchers of the United States, Canada, Australia, New Zealand, Argentina, and South Africa all helped supply the European market. However, the American farmer seems to have supplied half the market for cereals in Europe, and half of all European net imports came from the United States.

For centuries most Europeans had lived in great poverty in a subcontinent with fewer natural resources than most other comparable areas in the world. Europe had an abundance of mountains, streams, swamps, and other natural barriers to conquest and migration, but not enough first-class farmland. During the 19th century, Europeans began to reform age-old systems of class distinctions and strangling laws and institutions. Through industrialization at home and exploitation abroad, Europeans gradually improved their standard of living. Per capita food consumption rose: at first mostly of starchy foods, but later of meats, dairy products, and even delicacies. These economic and social changes were crucial to the success of the American farmer. Europeans of the 19th century not only needed the food, but they could pay for it.

For the American farmer, Europe provided the only important market outside the United States. As the 20th century progressed, Europeans had to rely more on themselves and on others for food. In spite of their great productivity, American farmers could not profitably supply both the domestic and the foreign market because of the increasing size of the rich American market. The American farmer

TABLE 13.3.   Approximate Population of the World and Its Subdivisions in Millions
          of Persons

Year	Europe	The Americas	World Total
1800	188	29	919
1850	266	59	1,163
1900	401	144	1,555

SOURCE: Bennett, *The World's Food.*

could not sell as profitably on the world market. The following population figures explain part of the change.

CHANGING CONSUMPTION AND POPULATION
PATTERNS IN THE UNITED STATES

But wealth was inequitably distributed, and urban and rural income
alike left few margins. Even so, Americans were far better off than
people anyplace else.

However, per capita consumption foods actually declined between
1899 and 1914 in the United States. Meat consumption fell from
150.7 pounds per person in 1899 to 140.0 pounds per person in 1914.
Potatoes declined far more sharply, from 187 pounds in 1909 to 157
pounds in 1914, and wheat flour fell from 217 pounds per capita in
1909 to 207 pounds per capita in 1914.

But the rapidly growing population made less food available for
export. Fresh beef exports dropped from 352 million pounds in 1901
to 6 million pounds in 1914, and bacon exports fell from 650 million
pounds in 1898 to 194 million pounds in 1914. Butter and cheese
exports fell from 79 million pounds in 1898 to 6 million pounds in
1914. In contrast, wheat and flour exports of 235 million bushels in
1901 only fell to 146 million bushels in 1914. On the whole, protein
foods declined most as exports and carbohydrates declined least. Production continued to rise in all commodities.

A look at the percentages of the various crops exported shows
that cotton and tobacco remained relatively constant from 1900 to
1914, but that wheat and corn fell sharply. In 1900 35.84 percent of
the wheat crop was exported; by 1913, this was down to 19.07 percent
of the total crop. The farmers least well-off were those engaged in
growing cotton and tobacco chiefly for export. The most prosperous
farmers exported the least, and the less they exported the more prosperous they were.

In spite of advances in technology and science and their applica-

TABLE 13.4. Annual Averages

Years	Gross National Product	Gross Flow of Goods to Consumers	Private and Public Capital Formation
1869–1873	6.71	5.38	1.34
1877–1881	9.18	7.53	1.86
1887–1891	12.3	9.58	2.69
1897–1901	16.8	12.9	3.89
1907–1911	30.4	24.1	6.35
1912–1916	38.9	30.8	8.05

SOURCE: *Historical Statistics,* 1957.

tion, food production could not keep pace with the needs and wants of Americans. Between 1900 and 1910, the American population increased 21 percent, but cereal production rose only 1.7 percent. The difference was made up out of what had been exported. Traditionally Americans had increased their food supply by opening new lands, but greater production per acre and per animal had to be the final solution for American food needs.

Tariff barriers against American products may have reduced some U.S. exports. However, the total population of the United States in 1860 stood at 31,513,000, of which only 20 percent was urban; the estimated population came to 99,118,000 by 1914, of which at least 48 percent was urban. The nonfarm population increasingly consumed more of the American domestic food supplies. At the same time, Europeans increased their imports from other parts of the world. But in the great southern commercial crop of cotton the Americans remained supreme.

Tremendous American industrial activity caused a constant population shift to cities in spite of a concurrent movement to western farmland during most of the 19th and 20th centuries. Industrial and agricultural activity, reflected in the figures for Gross National Product, show what happened. Table 13.4 gives five-year periods of annual averages in billions of current dollars.

FARM PROBLEMS AND EFFORTS AT REFORM

The war economy produced farm prosperity in the North and near ruin in the South. Northerners tried to remain in their old enterprises after the war and virtually forced southerners to maintain and even expand cotton growing. The South had to endure having the major profits from their enterprises siphoned off by northern businessmen.

Between 1867 and 1900, farmers put more land under cultivation than they had opened up between 1607 and 1867. The increase in farm acreage alone in 35 years showed changes which surpassed the changes of the total 260 years. Farmers of the East needed to increase farm size, to use more improved machinery, and to concentrate on crop and animal specialties in order to meet the demands of the rapidly increasing urban population. Of course, the largest increase in cultivated acreages took place in the West.

The most conspicuous characteristic of the changing times was the economic distress of most farmers. Between 1867 and 1898, they endured nearly continuous depression. Farmers suffered from their inability or unwillingness to adjust to changed circumstances. To solve their difficulties which were really troubles of the marketplace, farmers created new organizations to alter the market and marketing. The solutions they advocated usually would not have produced the desired results. Some were tried, others were not.

Beginning in 1867 and peaking about 1875, farmer protests took the form of the national organization called the Patrons of Husbandry or, more commonly, the Grange. Grangers appeared first in the Midwest, and then spread to New England, the South, and ultimately the Far West. They influenced the legislature of Illinois and some other states. They concentrated on control of railroads because they made commercial farming possible, but conversely made the farmers dependent on them. The Grangers secured state legislation setting rates and rules for the railroads. However, these regulations were put aside by the Supreme Court in 1885. The railroad regulations had not really worked well, largely because of the inflexibility of the laws. The alternative solution of regulatory commissions proved more helpful to the railroads than to the farmers.

But the Grangers did force on public officials the view that railroads were public utilities, subject to public regulation, which the Supreme Court upheld in 1876. The public utility concept was one of the most significant developments in American economics. It represented a median point between outright government ownership and unbridled free enterprise. The idea of public utility, which rejected both socialism and laissez-faire, was a uniquely American approach to the problem of responsible freedom.

INCREASED FEDERAL CONTROLS OF
BUSINESS AND FINANCE

Federal regulation and control of public utilities increased after Congress set up the Interstate Commerce Commission in 1887. Some of the old Grangers may have influenced the legislation, but the

Grange itself had little direct effect on the legislation. Instead, the Interstate Commerce Act of 1887 and the Sherman Anti-Trust Act of 1890 were passed during the ascendancy of the Greenbackers and the newly emerging Populists.

Even such regulation of rates and services as the Grangers had achieved failed to produce farm prosperity. In retrospect, the marketing problems were more complicated than simple exploitation by the railroads. Even with free transport of products, some commodities would still not have paid their cost of production.

In a complex market society, money is an intellectual concept with some material manifestations. Intrinsic value of money is not necessary for it to function or even exist in material form. Numbered papers—checks and bank records—can serve well enough. By the 1870s, the American business community was rapidly advancing to such levels of sophistication. However, farmers and workers had tied their business activity to the available amount and velocity of coin and paper currency. The urban business community, however, had a system of almost infinitely expandable money in the form of credits and checkbooks. Urban businessmen operated to some extent with symbols for money, while farmers and workers did not have the privilege. Expansion required a banking system of greater flexibility than existed in the mid-19th century. For the public good, such banking needed more public regulation.

INFLATION AND FARM PROBLEMS

Farmers saw their solution to the problem of market profits to be a simple one of increasing the amount of currency in circulation. Continuous inflation was advanced as the route to agrarian prosperity. This view had much historical experience to recommend it. From the Revolution on, inflation had been a source of farmer prosperity. From the legal tender of the Revolution to the bank notes of the wildcat state banks, the farmers had depended on "soft" money.

The Grangers directed their next major effort toward cooperative buying. Farmers often felt that middlemen made exorbitant profits. But farmers did not realize just what services middlemen performed and what these services cost. For a variety of reasons the cooperative enterprises did not flourish either. Business firms countered with better service and lower prices, thus drawing farmers away from their cooperatives. The Panic of 1873 which brought general economic distress also caused many of the cooperatives to fail. Many farmers could not afford to maintain membership in the Grange, and some members probably left farming altogether.

Many former Grangers, particularly in the Midwest, shifted over to the Greenback party. The Greenbackers appeared around 1874, ran a presidential candidate in 1876, and polled a million votes in the congressional elections of 1878. They ran presidential candidates in 1880 and again in 1884 when they disintegrated. Their program centered on printing more of the paper money which the government had issued during the Civil War. This money was being retired as early as 1866.

Not all the Greenbackers were farmers, or vice versa. The weakness of the party suggests that they offered an unworkable solution to farmers' money problems. The Greenbackers organized during a general depression, but before they could gather strength, the depression ended. Distress lingered on as men on small acreages, with too little capital and growing unprofitable commodities, still clung to farming. In these areas of maladjustment, men dreamed of success—if only there were more money available.

By the 1880s, railroad regulation, more currency, or control of middlemen did not appear to be easy solutions to profound farm problems. Several farm organizations, such as the Northern Farmers' Alliance and the Southern Alliance, were formed to work on marketing problems. Soon some labor unions joined in the general reform movement. In 1890, these groups tentatively formed the People's party which was organized officially as the Populists at Omaha, Nebraska, on July 4, 1892.

Instead of government regulation of public utilities (especially railroads), the Populists called for outright government ownership. They wanted free coinage of silver at an inflated price for gold, but deflated for silver—in effect, plenty of cheap money. They asked for more greenbacks. Believing that farmers could sell their produce at a higher price if urbanities could afford higher prices, the Populists also championed labor legislation to increase wages and decrease working hours. The whole program amounted to a general legislative approach to the problems of a market economy. Although the Populists elected some state officials, captured a few legislatures, and sent representatives and senators to Congress, they did not remain in control even where they won. Furthermore, they never succeeded in any urbanized area.

The panic and depression of 1893–1896 probably helped crush the movement. Contrary to popular opinion, real and deep distress tends to sap the enthusiasm of reformers and rebels. The Democratic party took over many of the planks of the Populist party in 1896, and the Populists faded from the American political scene.

CHAPTER FOURTEEN

TRANSPORTATION AND FOOD
PROCESSING DEVELOPMENTS ⚜ 1861–1914

T HE DEVELOPMENT of a national, rather than a regional, economy depended on moving raw materials to the factories and the manufactured products to the consumers. The operation of the world market lay outside any realistic hope of control, but the transportation system and the monetary machinery were at hand.

During the Civil War, American tonnage on the high seas declined and completed a trend which began in the late 1850s. Although most American overseas trade moved in foreign ships, the American shipbuilding industry did not decline and water transportation remained important to farmers. In fact, efficient water transportation was the most effective alternative to railroad monopolies.

In the post–Civil War period, shipbuilders gradually shifted from wood to iron and eventually to steel. The growing iron and steel industries, which supplied the metals cheaply for farm implements and machinery, also made them readily available for the shipbuilding industry.

Shipbuilding centered in those ports where iron ore (from the Mesabi Range of Minnesota or from Pennsylvania) converged with coal and limestone. The Delaware River ports and shipyards produced the most iron ships. Cleveland, where iron ore arrived by water and coke by rail, also became a shipbuilding center for the Great Lakes. Pittsburgh, in the heart of the iron industry, continued making riverboats. San Francisco, the best West Coast port, became a manufacturing center because iron and coal could come in by water. The flurry of shipbuilding activity in the 1870s did not, however, mean the end of wooden ships. They still predominated in tonnage built and in service.

Through the 1880s, the tonnage of steel ships rose from 31,000 tons per year to 124,000 tons per year. However, wooden ship tonnage built fell only from 401,000 to 320,000 tons for the same period.

TABLE 14.1. Comparative Grain Freight Charges, Railroad Rates, Chicago to New York (average rates)

Year	Wheat		Corn	
	Via Lake & Rail	Via All Rail	Via Lake & Rail	Via All Rail
		cents per bu		
1870	19.15	26.11	19.32	24.37
1875	12.71	20.89	11.34	19.50
1880	15.70	19.80	14.43	17.48
1885	9.02	13.20	8.01	12.32
1890	8.50	14.30	7.32	11.36
1895	6.95	11.89	6.40	10.29
1899	6.63	11.60	5.83	10.08

SOURCE: USDA Yearbook, 1899.

At the end of the century, Maine shipyards still turned out wooden ships in volume, but the timber came from outside the region. Steel replaced iron for shipbuilding rapidly after 1880 and made important increases only in ocean vessels with subsidies.

Congress gave the most important subsidies in an act of 1891 for iron or steel steamships which carried the mail. The law limited subsidies to ships of 8,000 tons or more, with a speed of 20 knots or more. The subsidy stimulated steel shipbuilding. The tonnage in coastal and Great Lakes trade increased and proved significant to farmers in their efforts to find markets. Transportation costs moved downward throughout the late 19th century. The savings, however, went to middlemen rather than to farmers. Comparative statistics for freight charges show what happened. The rates were reported by the New York Produce Exchange and by the Chicago Board of Trade and published in the U.S. Department of Agriculture Yearbook of 1899.

The big drop in the combined lake and rail charges took place after 1880. The decline corresponds roughly to the development of steel shipping on the Great Lakes with large tonnages and faster, more powerful ships. Improvements in railroads also had something to do with the declining costs of rail transport. The figures show average rail transportation costs also decreased.

EXPANSION OF THE RAILROADS

The miles of track owned and operated, including yards and siding, indicate some of the changes which occurred after the Civil War. A railroad boom took place in the West and the East. In 1860, the United States had 30,625 miles of track, mostly built before 1857.

Much destruction of southern lines and rolling stock occurred during the Civil War, while construction and rolling stock increased in the North. But by 1869, a train of standard gauge could cross the continent by selecting the right lines. Few odd gauges were left by 1886. However, most of the expansion of lines took place after the war. By 1870, U.S. mileage came to 52,922, despite losses in the South. Southern railroads made more substantial gains than in the years immediately before the war. North, South, and West, some 33,000 miles of new track were laid between 1867 and 1873.

The Panic of 1873 slowed construction awhile; but by 1875, the main eastern railroads—the New York Central, the Pennsylvania, the Erie, the Baltimore and Ohio, and the Grand Trunk—had emerged as efficient and relatively complete systems. Track mileage rose to 93,261 by 1880. At that point the nation was probably overbuilt. Americans had more rail service available than they needed in 1880. But by 1890, the miles of track had risen to 167,191, which meant over 70,000 miles were built between 1880 and 1890! By then the development of farming and industry had reached a stage where the rail service only slightly exceeded the needs of the country. By 1910, the operated rail lines came to 249,992 miles of track, mostly built during the "Golden Age" of American agriculture. By 1914, the United States had more rail mileage than all of Europe and more than one-third of all the track in the world. Commercial farming, which required adequate transportation, had become a reality in nearly every part of the United States.

IMPROVING THE HIGHWAY SYSTEM

As water and rail transportation improved, time in transit and costs fell. Agricultural produce could be shipped worldwide at low cost if the farmer could get his agricultural produce to the railroad or shipping dock. Road building methods improved little between 1861 and 1914. Roads were the responsibility of states and counties; cities took care of their streets. Most Americans traveled dirt tracks in country and in town. Even paved streets had inadequate or nonexistent drainage. Country roads had more open drainage, and wagons and charts had high, large wheels to keep the wagon bed clear of water and stumps.

Teams of horses pulling wagons of produce wallowed in mud in the spring. In the North, sleighs carried produce in the winter. Southerners, however, often suffered through sloppy roads half the year. In the eastern section of the country, the roads were usually

TABLE 14.2. Estimated Miles of Surfaced Roads, 1860–1914

Year	Miles	Percentage Increase
1860	88,296	
1870	92,265	3.3
1880	96,146	4.2
1890	106,200	11.4
1900	128,500	20.9
1914	257,292	100.2

SOURCE: *Historical Statistics*, 1945.

made of gravel, although brick and stone were sometimes used. The stone crusher and steamroller were both available in the 1850s to build macadamized roads.

Many improved roads were toll roads or turnpikes with tollbooths every few miles. Entrepreneurs collected enough to maintain their roads, pay employees, and accumulate some profit. Since the owners usually closed tollbooths at nightfall, many a farmer moved his produce at night to avoid the tolls. North and South, counties often used convicts to maintain roads, but farmers usually carried out simple maintenance of the roads near them, and they could work off part of their taxes. The unskilled labor took little of their time, and it usually occurred between seeding and harvesting when the farmers could spare some time from farm work.

The greatest strides in road building occurred after the automobile appeared, but the bicycle (built for one, two, or more) took the country by storm in the 1890s. The American Wheelmen, a national association of cycling zealots, led the way in demanding better roads. Under the leadership of bicycle manufacturers, particularly A. A. Pope, the states undertook expanded road building programs, which incidentally benefitted farmers. Improved roads constructed between 1890 and 1914 rose dramatically.

By 1914, gravel and stone roads were fairly numerous in the East, especially in the Middle Atlantic section. The farmer could travel about thirty miles round trip in a day. The typical open wagon held 50 bushels of wheat. Cattle, sheep, and hogs were usually trailed to market well into the 20th century.

By 1896, the first cars appeared in the United States, but they were expensive novelties until Henry Ford began mass production of his Model T in 1909. He first used an assembly line in 1913. The automobile industry did not exist in 1900; by 1914, it had already become the eighth largest in the United States. Table 14.3 shows the dramatic increase in the use of trucks and cars in the years 1910–1914.

TABLE 14.3. Trucks and Cars on Farms, 1910–14

Year	Automobiles	Trucks	Tractors
1910	50,000		1,000
1911	100,000	2,000	4,000
1912	175,000	5,000	8,000
1913	258,000	10,000	14,000
1914	343,000	15,000	17,000

SOURCE: *Historical Statistics,* 1957.

CHANGES IN MARKETING

Distribution of agricultural commodities changed between 1861 and and 1914, in response to the growth of cities, population, and territory in the marketing system. Formerly, merchants had had to personally tend to the larger markets in the cities. The railroad and the automobile enabled larger merchants, wholesalers, and importers to send out representatives seeking outlets instead of letting the outlets seek the suppliers. Specialization increased and distribution became more complex and demanded more expertise to be effective and economical.

The marketing agents expanded their activities. Commission agents usually handled commodities in carload lots and dealt in such items as butter, cheese, fresh fruits, and vegetables, but not in cotton, wheat, or livestock. The jobber worked with less than carload lots and, like the commission agent, used a warehouse for his activities. In cities, the larger warehousemen often served the retailers as well as smaller warehousemen. Dairy and vegetable farmers usually dealt directly with commission men and jobbers. Increasingly, the dairy farmers developed cooperatives to wholesale and retail their commodities. Cooperatives also tried to establish brand names for their goods for a national or regional market.

Efficiency in transportation, handling, and storing made the development of national marketing possible. Processors developed distinctive packages and sold under product brand names—for example, "Jersey Rolls" (margarine) in the 1880s. The National Biscuit Company and Uneeda Crackers led the way for other industries in packaging and in advertising. Advertising helped create national markets for some types of farm produce. The invention of the rotary printing press, sulphite paper, and later the linotype technically made display advertising possible on a large scale. Some 4 percent of the national income went for advertising by 1910. The farmers did not necessarily share in the economic benefits of the expanding marketing machinery. Still, the Golden Age of American agriculture (1898–1914) did coincide with the rise of brand name products and advertising in mass media.

RISE OF LARGE COMMODITY EXCHANGES

The development of large commodity exchanges began with the New York Produce Exchange (1862) and the cotton exchanges of New Orleans and New York (1871). Similar exchanges for agricultural products soon appeared in other cities. Most of the exchanges began as cash markets with the goods sold at auction, and many remained essentially "spot" markets. However, futures buying and selling soon developed in many exchanges, especially in the grain markets. The produce exchange could flourish only when demand for the product extended across the year without much change, and the product could be graded and given economical long-term storage.

The commodity exchanges produced a general stability in market prices. Some buyers and sellers, especially if dealing in volume, chose the exchanges over any direct dealing with the producers. Price stability favored the producers, but the economic power of the marketing institutions also was increased. The farmer had less recourse to the local market, which tended to follow the trends of the national or regional exchanges. Thus the wheat farmer in Kansas came to be in the power of not only the Kansas City or the Chicago market, but of even more distant exchanges. The exchanges averaged conspicuously similar prices. A famine in one area was easily overcome by a surplus somewhere else; a shortage of goods would benefit farmers in terms of prices only on an average. Local prices advanced or declined little.

Trading in futures and in storable commodities permitted and even encouraged speculation. The speculator performed a useful function as the key agent in price stabilization.

The mass market encouraged millers to concentrate on a single high quality of flour. The marketing and the processing called for better distribution. In the 1870s, the new process flour mills could grind as many as 4,000 to 5,000 bushels of wheat a day; two such mills ground in a day the wheat grown on 225,000 acres. The commodity exchange and the intermediate enterprises were necessary to keep the mills supplied with nine, ten, or more carloads of wheat in a day. Systems of purchase, storage, and transport had to be developed.

New mass processing and selling methods left many of the old-style wholesalers, brokers, and jobbers behind. The number of individual firms declined, but the workers in the system increased, particularly salesmen, advertisers, and other personal service representatives.

At the terminal end of the marketing chain, the old system continued well into the 20th century. In the handling of packaged products or small lots, such as vegetables, little change took place, and human lifting and lugging moved most commodities. Workers had aids such as hand trucks, pulley systems, gravity conveyors (usually rollers of some sort), and elevators. But to a surprising extent products moved from railroad cars to motortrucks and wagons by human strength and skill.

DEVELOPMENT OF THE CANNING INDUSTRY

The development of canning opened up a market for vast quantities of vegetables and fruits which led to specialized development of truck farming and the gradual elimination of the casual market gardener. The market demands also stimulated the improvement of methods of food preservation.

Early canners had learned to place filled cans in boiling water and heat them to 212° F. In 1861, Isaac Solomon of Baltimore added salt to the water which could then be heated to 240° F. This allowed canners to reduce processing time from five hours to half an hour. Without increasing plant capacity, a canner could increase his production ten times. In 1874, A. K. Shriver, also of Baltimore, invented the autoclave which processed the cans with live steam. Its use further increased production by cutting the processing time and reducing the danger of spoilage.

Techniques of can making improved from the hand-cutting methods of the tinsmith to the automatic soldering machine. By 1876, the conveyor and soldering machines enabled two men to produce about 1,500 cans a day, thirty times more than by the hand method. Before 1880, the cans had had small holes in the top through which small pieces of food were pushed. The open top can was invented in Europe. It speeded the business of filling the cans and allowed the tops and bottoms of the cans to be crimped on by machine instead of by soldering. Canners resisted crimping until after 1897 when new machinery was invented. The early crimping devices did not produce uniformly tight seals. Inventors made other improvements in canning machines until a single set of machines could make 35,000 cans a day by 1910.

Production of canned goods rose far more rapidly than the population. In 1860, canners put up about a million cases of food; by 1914, they canned around 90 million cases. The farmer as producer felt the impact of the new industry in more stable prices, a certain market, and a gradually increasing dependence on the market and on a few processors. Commercial canning of perishable commodities preserved food so that it could be easily handled and transported.

Ways of preserving meat included drying, smoking, canning, and pickling, but fresh meat could be moved only on the hoof. The railroad made possible transport of live animals but live meat still had to be fed and watered at extra expense. If the hides were needed in one place and the meat in another, the products had to be separated and reshipped at additional expense. Shipping parts of the slaughtered animals was far more economical; however, difficulties of preservation in transport presented the use of this method before the Civil War.

IMPROVEMENTS IN REFRIGERATION

The different railroad track gauges, the incomplete lines, and the resulting transfers made refrigerator cars for long-distance hauling impracticable for meats and other fresh foods although artificial ice could be made. In the early 1860s, engineers devised methods of mounting car wheels so that the wheels could be moved to conform to the tracks. In January 1864, the first freight car ran from Chicago to Boston on its own wheels without unloading the freight, and the refrigerator car finally became possible. Later in 1864, the Star Union Transportation Company began the fast movement of grain and other bulk commodities from Chicago to the East. The shifting of wheels to match the tracks remained the chief technological advance for some years.

The Star Union Transportation Company pioneered in sending meat and other perishable commodities East in rolling iceboxes in 1865. J. B. Sutherland of Detroit made a car with adequate insulation, ventilation, air circulation, and ice storage to provide useful and reliable refrigeration. Patented in 1867, his car applied the knowledge of refrigeration to the railroad car. As railroads gauges were standardized, lines of track were linked up, and the whole system of long-distance rail transport became more economical. Still, the railroads themselves did not invest in refrigerator cars.

What the Star Union Company began, Gustavus Swift continued. As a meat packer in Chicago, he wanted a cheap, effective way of getting fresh meat to the East. In the 1870s, he invested heavily in his own refrigerator cars which the railroads would haul for a fee. Soon Armour and other packers had their own fleets of refrigerator cars.

In 1869, Washington Porter brought a refrigerated shipment of bananas from California to Chicago. In 1870, he successfully used refrigerator cars to bring fruits, vegetables, and salmon from the Pacific Coast. Products of warm climates could now be produced for the off-season trade at a profit. Vegetables and fruit farming expanded beyond the areas near urban centers. Fresh products of this sort moved mostly in a limited luxury urban market, but the simultaneous development of the canning industry absorbed the surpluses. The grower of vegetables and fruits could profit consistently only if he produced in large volume. The small garden farm gave way to the ever larger complex worked by underpaid migrant workers. This development was already under way by the 1880s.

Shipping of refrigerated products solved only part of the problem of reaching the consumer. After the products arrived at a terminal someone had to transship them through the complicated machinery of warehouses, retailers, commission merchants, and jobbers. Cold-storage houses and retailer refrigerators had to be developed. Large iceboxes sufficed in the 1850s when few merchants dealt in refrigerated goods. After the Civil War, however, the system became completely inadequate. Europeans had developed the technological improvements necessary for the industry. Blockade-runners had carried two European mechanical refrigerators through the northern blockade into New Orleans early in the Civil War. In later years, Americans imported other plants from Europe. In the 1880s, Americans began large-scale production of mechanical refrigerators. By 1900, the number of mechanically refrigerated cold storage warehouses had increased dramatically, especially in the smaller towns.

The appearance of the machinery of refrigeration from warehouses to railroad cars resulted in efforts to transport and store frozen foods. In 1914, merchants froze and successfully shipped barrels of cherries, berries, and other commodities. However, freezing broke down cells and destroyed color and affected the taste of the product. Quick-freezing appeared only after World War I.

OTHER METHODS OF PRESERVING FOOD

In the 1860s and 1870s, great bacteriologists such as Pasteur and Koch did their pioneering work and discovered the microbal origin of many diseases. They found various methods, some chemical, to combat microbes. Independently of the discovery of bacteria, European medical scientists had discovered the usefulness of germicides and antiseptics. The discovery that bacteria caused food to rot stimulated processors to add quantities of various adulterants such as carbolic acid, boric acid, benzoates, and other chemicals to retard or halt deterioration. Food was nearly embalmed in some cases.

From the 1880s until the passage of the Pure Food and Drug Act of 1906, processors preserved food by using increasingly potent adulterants and "unnatural" additives. Whatever their baneful effects on consumers (poisoned meat was fed to soldiers of the Spanish-American War), the preservatives did open up new possibilities in transport of farm commodities. A quarter century of uncontrolled food preservation may have paved the way for the development of farming and ranching on a scale of profitability otherwise impossible.

RISE OF CHAIN FOOD STORES

Retail outlets had always abounded, but the rise of first the chain store and then the supermarket brought many changes. The A&P (Atlantic and Pacific) chain had already appeared by the time of the Civil War. Other chains appeared later—the Kroger Company in 1882, the National Tea Company in 1899, etc. A & P began self-service of the American supermarket type in 1913. The retailers sought to have the customer do the work of clerks. This method not only cost less, but it increased sales volume. The consumer paid lower prices, achieved higher levels of consumption, and created a steady, reliable market for farmers.

In many ways, the key to American farming was the development of efficient transport, from shipping on the high seas to sacks of groceries on the back seat of the automobile. In economic terms, transport included not only ships, boats, trains, and trucks, but also elevators, warehouses, refrigerators, cans, boxes, and sacks. Food processing developed chiefly as an adjunct to transport.

CHAPTER FIFTEEN

TECHNOLOGY ON THE FARM:
MECHANICS AND BIOCHEMISTRY ⚹ 1865–1914

T ECHNOLOGICAL and scientific change accelerated in farming and industry between 1865 and 1914. Until 1914, the farming areas constantly expanded as frontier areas came into production. The world market could absorb the surplus products of the farm. However, the percentage of those in farming fell steadily. By census years the percentage in agriculture was: 1870, 53.0 percent; 1880, 49.4 percent; 1890, 42.6 percent; 1900, 37.5 percent; 1910, 31.0 percent. Farmers became a minority of the work force sometime between 1870 and 1880, although their numbers rose absolutely until the 1930s. The number of farmers growing the five major crops, including cotton, declined more than those in livestock enterprises, including dairying. Animal husbandry still required comparatively large amounts of human labor. Advances in crop husbandry did not make much impression on animal husbandry since farmers grazed many animals.

In commercial agriculture especially, time was money and production in volume was essential. The farmers sought improved plows that could do the work quickly with a minimum of power. If they could keep labor expenses low and work animal numbers down, they could get more profits. If 100 plowmen did the work on a bonanza farm, profits would rise if the number of plowmen could be cut to 50 or less. The commercial farmer's major concern was to provide food quickly, cheaply, and on a large scale and thereby make a handsome living.

NEW AND IMPROVED PLOWS

Until about 1900, plowing was generally done with animal power. Although the disk plow worked well, especially in light soils, it found little favor. It required about 15 percent more pull on the drawbar than the moldboard plow. The disk plow could not be used economically enough with horses or oxen. Disk plows became practical when tractors came into use. Farmers particularly wanted a plow that would scour well; that is, break the soil while leaving the plow, share, and moldboard free of dirt and mud. The slicker the plow surface, the faster animals could pull it and the less it would clog.

In 1868, John Lane invented a comparatively expensive and effective plow material, called "soft center" steel, and made an iron plow with thin layers of hard steel on the sides. Farmers preferred the plow for virgin land and hard soils if they could afford it.

James Oliver, a blacksmith, developed a cheaper plow material which performed well under adverse conditions. He patented a method of hardening cast iron in 1868 and brought out his commercial model in volume in 1870. He sold his plows everywhere, including the South and New England. He seems to have modeled his plow after an earlier plow known as the Michigan plow. Other manufacturers continued to make and sell plows, and Jethro Wood's Eagle plow remained in use in one form or another for many years.

The new plows reduced the draft required to do the work effectively. Although the plows moved through the soil faster, they could not be raced across the fields. Workhorses, three to a plow, could do about two acres in a day with the one-bottom walking plow. Even at this slow speed, the plowman had power to spare. Two uses could be made of the extra power. First, the extra draft could be used to carry the plowman with no added strain on the animals which led to the development of the sulky plow, or one-share riding plow. Second, the extra draft could be used to draw more than one bottom with no reduction in speed. This also required a sulky arrangement since the farmer could not walk behind and manhandle a large gang plow.

Serious efforts to build sulky plows took place in the early 19th century. But the sulky plow on a small commercial scale was not introduced until about 1864. About 26 plows were sold that year, and several types and brands were available by 1868. The sulky took hold rapidly on the Pacific Coast, to some extent in the East, later in the 1870s in the Midwest, and then the South. Suppliers could not keep up with the demand from 1876 to 1878. By the 1880s, sulkies were commonplace from Ohio to Iowa. The three-wheeled sulky plow appeared in 1884. The wheel replaced the landside and further reduced the draft. Various brands were on the market, including some by John Deere. The East and the South still lagged behind in the use of sulkies, but the rest of the country had generally adopted them by 1890.

Another way to use the excess draft power was to increase the number of bottoms, that is, make a gang plow. The term "gang plow"

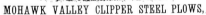
Poster, farm machinery, *ca.* 1865.

had been used for some time to describe a walking shovel plow. These plows were really forms of cultivators that could be used for cross plowing or to break very light soil. They cast no furrow and could not be spaced very far apart before they became too unwieldy for a walking farmer to manipulate.

The sulky gang plow began to appear in the upper Mississippi valley in the late 1860s. Earlier versions had a heavy draft which could be improved by a smooth-scouring, quick-cutting plowshare and moldboard. But by the 1870s, after Lane and Oliver, the sulky gang plow increased in use. Farmers especially used them on the Pacific Coast and in the Red River valley of the North, both wheat bonanza areas.

Most gang plows were only two-bottom plows. More draft required more animals in most soils and made the additional furrows less economical. Farmers normally used four horses, although from three to eight could be used. By the 1890s, the larger commercial grain farmers commonly used eight horses to pull four bottoms. Iowa farmers of the 1870s called the gang plows "horse killers." They cut back to one-bottom riding plows.

Historians have slightly misdirected the credit given to plow makers, such as Deere, Lane, and Oliver, in speeding the settlement of the prairie. The time saved by farmers who used the clean scouring plows did not equal the savings produced by the gang plow. The effective gang plow depended on the prior invention of steel shares and moldboards. The sulky gang plow doubled the amount of land a farmer could handle. Therein lay the rapid conquest of the prairie.

IMPROVEMENTS IN HARROWS, SEEDING EQUIPMENT, AND CULTIVATORS

In all sections of the country, farmers generally plowed and then harrowed as separate operations. The harrow "A" frame or box frame differed little from those used in the 18th century. They made the harrow of strong materials and hinged the sections to cross uneven land. In the mid-19th century, possibly from Japanese or European examples, Americans began experimenting with rotary or disk harrows.

Shortly after the Civil War, a form of disk harrow, the Nishwitz harrow, came into use. In the 1870s, several types of disk harrows were made in Massachusetts and New York. Eastern farmers were the chief customers, but the makers advertised their harrows nationally. In 1880, the Keystone Manufacturing Co. of Sterling, Illinois, began making disk harrows. Their production and use rose appreciably after

1892. The disk harrow could cut and turn under stubble, and it further broke up heavy soils. It did not work well on rocky soil and it was not needed on lightly packed soils. Under those conditions, farmers preferred the spring tooth harrow. It cleared rocks and roots without breaking and also cleaned itself. The spring tooth harrow seemed especially suited to well-cleared farmlands tending to rocky soil. For covering seed after planting, the spring tooth harrow did the job well with less draft than the disk.

In 1869, David L. Garver patented the spring tooth harrow, using teeth of spring steel. Other patents improved the device, especially in 1877, and it soon came into widespread use in the East and the South. By 1878, manufacturers in New York and Ohio made the spring tooth harrow, and most Ohio farmers had adopted it by 1883. Generally, the spring tooth harrow found favor in the East: in New York, New England, and the Middle Atlantic region down into Virginia.

Broadcast sowers, on carts (endgate seeders) or carried by the sower, continued in use, mostly in the Atlantic seaboard states. The seed drill of the 1860s still found little favor because it did not work well enough on rough or rocky land. Furthermore, growers did not need it in the wheat areas of the Midwest and Pacific Coast because of the fertile soil. Those farmers did not pulverize their land in the plowing and harrowing processes. Production results justified the continued use of endgate seeders in the Northwest and on the Pacific Coast as late as the 1870s.

The seed drill became more important to farmers of the upper Mississippi and the Far West. Better plows and harrows made the careful and complete preparation of a large seedbed possible and economical which was important as soil fertility lessened. The relationship between rapid, thorough plowing and harrowing and the increased popularity of the grain drill is direct and obvious.

The drill had the great advantage of economical seeding which commercial wheat farmers found especially important in times of price declines. By 1874, farmers in the wheat belts rapidly adopted an improved drill that operated well in mud or over an occasional root or rock. The farmers favored the Buckeye and the Farmer's Friend brands, but all types were improved and sold. The feeding mechanisms were improved to release the seed without waste. Machines got progressively larger because the more rows seeded, the greater the utility of the machine for the commercial grain farmers.

Various furrow openers had been used since the 18th century— from coulters, to chisels, to shovels. The shoe drill appeared in 1855. By the 1890s, the disk had been found to be the best in most soil because it penetrated the soil well without clogging. By 1890, most farmers used the well-designed and inexpensive drills if they had fairly level land without many obstructions. The ever-larger areas seeded

Poster, farm tools and equipment, *ca.* 1875.

Cultivator, first known gasoline-powered, 1912.

were made possible by simultaneous improvements in plows and harvesters, which also stimulated the use of the grain drill.

The basic corn planter had been invented in the 1850s. A check-row device operating off twine knots at intervals became very popular after its mass appearance in the 1870s. In the drier portions of the country—the eastern edge of the Great Plains and the western edge of the prairies—corn growers planted deeper because of insufficient top moisture. To meet their needs, a combined plow and planter was developed about 1880. A plow with a double moldboard, called a lister, which cut deeply and threw dirt in both directions, left a series of trenches in the field. The seed fell into the trench from an attached box, and a small roller behind covered the seed, all in one pass. This device was used past 1914, but only where the climate demanded its use because of the heavier draft required.

In corn country, plowing and planting remained as two distinct operations through 1914. Corn culture and harvesting demanded more attention and time than other grains. The farmer could bear the time and expense because of its fantastic seed:yield ratio—about 1:150!

Cultivators for corn, cotton, and other row crops were well developed before World War I. The improvements before the turn of the century included adding a seat for the farmer, making the blades spring clear of obstructions, and devising ways of raising the hoes for turning around at the end of rows. No one used more than two-row cultivators, and many farmers preferred one-row implements. Light shovel plows predominated in cotton culture, and the hand hoe was mostly used to chop or cut a stand. Inventors attached hoes on drums to chop cotton to a stand, and planters sometimes used them. Although potato planters were available by 1856, they came into general use only in the 1880s. In the 1890s, farmers often used newly invented machines for transplanting vegetables and tobacco.

INCREASING USE OF FERTILIZER AND ITS APPLICATION

In general, the longer an area had been farmed, the more fertilizer farmers had to use to grow a profitable crop. Although they used some commercial fertilizers, such as lime and guano, they chiefly fertilized with animal manure and sometimes plant manure. The manure was spread from wagons and distributed with forks on the fields. Eastern farmers first made widespread use of patented manure spreaders. Most of the spreaders worked from the endgate in various ways. In the western grain areas in the 1860s and 1870s, growers used little manure because they had few animals and large acreages. Animal bones provided the chief source of phosphorus. Professional hunters systematically killed the buffalo of the Great Plains in the 1870s, and some men collected the bones and shipped them East for use as fertilizer. The bones were ground and then retailed in sacks. Sometimes seed drills had attachments for spreading this processed fertilizer.

Endgate fertilizer spreaders continued in use in the Mississippi valley, the prairies, and the Great Plains during the 1860s and 1870s. In contrast, eastern farmers commonly used fertilizer attachments on drills by the 1870s, if they used drills. All areas seem to have used a variety of the endgate spreader which appeared about 1865. In 1877, commercial production of a traveling apron type of wagon spreader began. This kind of endgate dumper was used for manure as late as 1914.

Farmers used increasing amounts of manures and fertilizers between 1865 and 1914, especially in the upper Mississippi valley, the Middle Atlantic and the New England states. The farmers could afford it and had a ready supply. The cotton lands of the South, with their depleted soils and gully erosion in many places, needed soil restoration and fertility preservation. But the cotton South was poor. The sharecropping system punished both landlords and workers, and fertilizer could not be applied in the amounts needed. Only the remarkably fertile soils of the Black Belt and the Mississippi flood plain continued to yield well.

Southerners had few manure-producing animals, especially in the lower South. The relative lack of dairying no doubt contributed to the problem. In the corn-hog-bourbon areas of the highlands, hogs might have produced sufficient quantities of manure. Other farming practices, such as shallow plowing and up-and-down plowing on hillsides, made any application of manure less effective.

The Middle Atlantic and the New England states had one great advantage in the use of manures and commercial fertilizers. Their large cities encouraged commercial dairying, which rose to new importance from 1865 to 1914. Throughout this period the chief dairy state was New York, followed by other eastern states until later in the period. Wisconsin dairying did not become significant until the 1890s, and Wisconsin only emerged as the leading dairy state in the 1920 census.

Dairying was one of the most profitable farm enterprises. It provided manure for fertilizer and sometimes extra to sell. Those involved in livestock or poultry husbandry made the most use of manure. Insofar as they did, they also achieved better yields and higher returns, and these enabled them to buy other chemicals needed for the soil. The rise of truck farming in the Northeast also contributed to the greater use of manures and fertilizers, as did the ever-flourishing livestock feeding industry. In the 1870s and 1880s, feeders located near the urban centers.

Truck farms, although intensively farmed and thus rapidly depleted, flourished because their comparatively small acreages required smaller amounts of fertilizer. The farmer on such profitable land could buy the fertilizer he needed, and farmers did trade in manure. Livestock feeding was often pursued in conjunction with something else. Feeding produced manure—sometimes more than the farmer needed for his own fields. Grassland farming, already practiced in the animal specialty areas of the East, required little manure as practiced at the time. The stockman kept his land in shape by natural manuring and by planting clover and other legumes.

Dairying, truck farming, and feeding livestock saved the Northeast and the Midwest from the devastating ravages of erosion by wind or water. Most technological advances in U.S. farming increased the rate of exploitation of resources. Better plows allowed more rapid and efficient plowing, opening to erosion otherwise protected land. More effective seeding and harvesting produced a more rapid depletion of fertility. In contrast, many advances in dairying and other commercial specialties tended to decrease exploitation of the land while increasing productivity.

USE OF SILAGE AS LIVESTOCK FEED

Of these advances, few had such widespread ramifications as the discovery of silage. The first silo was probably built in the United States in 1873. Grass, grains, legumes, and other plant stalks could be stored for prolonged periods without spoilage with a minimum of labor. The nutritional value and palatability increased because of the biochemical reactions inside the silo. Ensilage markedly reduced the dangers of bloating for animals fed on alfalfa and clover. Sorghums, introduced as important feed plants in the West in the 1870s, poisoned animals unless the plants were cured as hay or silage. The stalks of the sorghums used for sugar in the North and East were made into livestock feed by ensilage.

Because of the increased feeding value of silage over hay, farmers could carry more cattle over winter than previously. In addition, the silo greatly reduced the storage space needed for feed. It was easier to fill than a barn. Figures on silos do not exist for the early period. The buildings and the method spread slowly at first. Journalists and scientists urged its adoption, and the method was soon widespread. By the 1890s, most dairymen used the silo; by 1900, it had spread to peripheral dairy regions; and by 1914, some feeding areas had silos as well.

ADVANCES IN DAIRYING

With full winter feeding, the annual production of cattle rose steadily and the price fluctuations between summer and winter lessened. Death losses in winter and in the early spring declined. As farmers found they could take care of more cattle, they acquired larger herds and more farmland. Between 1880 and 1914, the average size of dairy farms grew, and the number of cattle on farms rose. Between 1865 and 1914, farm editors and others kept reminding farmers that cows treated with kindness yielded more milk. The rise of productivity associated with changing attitudes cannot be catalogued, but the development of the silo and the increasing use of stables during the winter helped to decrease cruelty and to increase productivity.

Although some farmers had purebred dairy cattle, most dairymen kept grades or crossbreeds, and generally they did not keep the best hybrids. The movement to purebred herds of Holstein-Friesians, Jerseys, Guernseys, Alderneys, Brown Swiss, and Ayrshires took place after further advances in science.

The dairy farmer's most valuable commodity was butter because it brought the best returns and could be produced at some distance from an urban market. It helped set the prices of other dairy products. (Oleomargarine entered the American market around 1873 and posed no real threat to butter until after 1914, although dairymen vigorously attacked it from 1880 onward.) Dairymen had no quick, accurate way of determining the butterfat content of milk. Nor did they have any sure way of telling which cows gave milk rich in butterfat. Waiting for the cream to rise in a glass of milk and then estimating the percentage of cream to the whole produced inaccurate results of little value.

Stephen M. Babcock invented a comparatively simple butterfat test in 1890 and put it on the market in 1891. The test employed acid, heat, and centrifugal force to precipitate a measurable percentage of butterfat. The method had wide-ranging effects. Butter factories and others could pay according to the amount of butterfat. Payment differentials favoring the higher testing milk gave economic impetus to farmers to increase the fat content of their product. Although many resisted the payment scheme, the butter manufacturers insisted and prevailed. So the commercial farmer needed to rapidly and accurately identify his best producers. The Babcock tester could not always be used successfully by the farmer (an amateur scientist at best), but he could have his milk tested at small expense. Associations for the testing of cattle soon appeared in dairy country. Large-scale testing was just getting under way by 1914, but it was already important to the industry. The butterfat tester not only identified the less productive cows called "boarders," but it also helped farmers identify the feed and care which led to the highest butterfat production.

The use of silos got an additional boost, based on scientific evidence as a result of the Babcock test. Silage proved to be the feed which produced the most butterfat. Scientists had suspected this anyhow, but substantial evidence proved the advantages of the silo. Furthermore, dairymen could improve the whole complex of the rations to increase butterfat yields. In general, care and feeding improved as the best cows were identified through the Babcock test.

Another discovery concerned the superiority of purebred cattle. The farmer could easily identify his best producers and sell off or slaughter the losers. For the most part, the purebreds produced best. The most successful dairymen had purebred or nearly purebred herds, and dairymen increasingly used good crosses of the purebreds.

As the search for good butterfat producers went on, farmers also increased the gross milk production of their animals. Farms tended to become larger as the business became more profitable. The growing urban population produced a continually expanding market for dairymen. In terms of cash income, dairying rapidly became a major industry and was the leading commercial farm enterprise for awhile in the 20th century.

Generally, the dairy husbandry complex flourished near major urban centers and along well-developed transportation routes. The areas nearest the cities tended to concentrate on fluid milk, more distant areas on butter, and the fringe areas on cheese. By 1914, the urban areas had been fairly well established. As the industry became more efficient technologically, part-time dairying became less profitable. The overall region of dairying shrank as the business became more productive in animal efficiency, in yields per acre, and in yields per man-hour. The first efficient milking machine and other dairy devices came into wide use after 1914.

The increased commercialization of dairying aided the commercialization of other farm enterprises. The virtual end of wheat growing in Wisconsin and other dairying areas illustrates the change. Wheat farming extended to the outer edges of metropolitan areas. Wheat growers were concentrated geographically because of transformations in wheat farming and because competitors in other places left the crop for regionally more profitable enterprises. Livestock feeders tended also to be located fairly close to cities and thus in the vicinity of dairying and truck farming. Having related interests, the industries could borrow techniques and could trade among themselves.

From the appearance of the hand-cranked centrifugal cream separator in 1884 to about 1905, dairymen often fed their skimmed milk to swine. They thus competed in the feeding industry. Home skimming of milk fell into disuse, although it continued on a minor scale until the 1970s in some places. The skim milk became a byproduct of butter factories which either sold it to consumers or to livestock feeders. Methods of drying milk were developed fully in the mid-20th century and practically took it out of the feeding industry altogether.

Poultry producers and livestock feeders, however, could adapt some of the techniques first used for dairying. Feeders increasingly used silage, and they all needed corn and other feed grains. They also implemented advances in automation of cattle feeding. The advantages of stall handling of cows also applied to poultry and swine husbandry. New developments included the first American commercially made incubator in 1885, a litter carrier and a feed carrier in 1897, steel stanchions for dairy cows in 1902, and a colony type of heated brooder for baby chicks in 1910. The increasing costs of specialized devices tended to separate the dairy and poultry industries. Poultry raising had been an adjunct to the dairy industry previously.

Better animals and better feeding methods required more skilled labor. Until the milking machine was developed, dairymen needed more hands to milk the cows and do other unmechanized chores. In the long run, science and technology produced savings for the livestock and poultry raiser. But the expenses for purebred stock and highly specialized items—such as the silo and the Babcock tester—made it unprofitable for the small or casual operator to enter or stay in the business.

Commercialized, specialized agriculture became necessary for the farmer's survival. Few farmers could afford specialized dairy equipment, specialized field equipment, and specialized poultry and gardening equipment. Farmers received their best returns by concentrating on one enterprise.

ADVANCES IN BREEDING AND DISEASE CONTROL

Success in either plant or animal husbandry ultimately depended on using the best living materials for human manipulation and regulation. In 1865, Gregor Mendel laid the groundwork for genetics; in 1900, Hugo de Vries enunciated a workable theory of genetic mutation. American scientists and others tried to find the best plants and animals by importing useful strains—from dairy and beef cattle, about 1817 to 1869; wheats, ranging from Turkey Red, brought in by Russian-German Mennonites in 1873, to the many varieties of hard wheat introduced by plant explorers such as Mark Carleton; to Egyptian cottons in the late 19th century. The constant improvement of strains by introductions from abroad and experimentation at home provided the American farmer with better plants and animals which yielded more abundantly.

The discovery of disease-causing microbes in the 1860s and of ways of combatting them meant better control of animal diseases. Higher economic returns resulted from cutting losses from death or disability of animals. Disease control called also for specialized equipment and knowledge which favored the highly specialized commercial farmer.

Many diseases of animals, such as tick fever, pleuropneumonia, tuberculosis, and hog cholera, were attacked by scientists from 1865 to 1914. The mosquito origin of malaria and yellow fever became known and corrective measures were taken. The need for sterile care in treating the sick was established beyond question and many advances in surgery were made. Knowledge of immunization opened new opportunities in veterinary as well as human medicine.

An explosion of knowledge in the field of biochemistry occurred in the late 19th century. In 1882, Robert Koch isolated the microbe which caused tuberculosis; in 1889, Theobald Smith and others proved that ticks carried Texas cattle fever. In 1890, Koch devised the tuberculin test for diagnosing bovine and other forms of tuberculosis. In 1891, the Federal Government began free distribution of a vaccine for blackleg. In 1892, the government eradicated the last case of pleuropneumonia in cattle by slaughter. In 1903, Marion

Dorset developed a vaccine for hog cholera. The end of pleuropneumonia by slaughter proved what could be done in eradication of diseases. Tuberculosis and hog cholera would not be brought under firm control until during World War II, but the method had been discovered. Before 1914, tick fever was nearly controlled and blackleg was partially controlled. The savings produced by advances in bacteriology amounted to billions between 1890 and 1914.

Plants suffered from diseases and from insect attacks, much as did animals. The cattle dip for ticks was paralleled by the spraying of insecticides and fungicides on plants. In the same year that Koch found the tubercle bacillus, another scientist discovered the Bordeaux mixture for killing fungus on vegetables and fruits. At first, a hand pump sprayer was used; but as early as 1887, a spraying device, working off a ground wheel, was used in the United States.

Insects presented special problems because any agent that could kill insects was usually harmful to humans as well. Truly effective and selective insecticides came only in the mid-20th century. Scientists also tried using natural enemies to combat insects with limited success. The destructive citrus fruit disease, fluted scale, appeared in California around 1868 and threatened the industry with extinction. In 1888, C. V. Riley, Department of Agriculture entomologist, sent Albert Koebele to Australia to search for a natural enemy of the fluted scale insect. Koebele returned in 1889 with ladybird beetles which devoured the parasites and saved the industry. Scientists tried the same approach on grasshoppers and other insects without the same success. However, agricultural biochemistry in all its forms was still in its infancy. The microbiological causes and several types of carriers of plant and animal diseases were the major discoveries of the years 1865 to 1914.

CHAPTER SIXTEEN

GATHERING IN THE BOUNTY:
HARVESTING ⚒ 1865–1914

HISTORIANS of technology view history in terms of mechanical or scientific changes. If a crop remained important but still was not influenced much by technology, it might go unmentioned for half a century because no great changes took place in its harvesting methods. Some crops, such as tobacco and cotton, saw few harvesting changes. Orchard products also increased in prominence in spite of few changes in harvesting. Such examples contradict the idea that production and economic viability are tied inevitably to advances in harvesting.

Historians have speculated that slave labor made technological advance unnecessary so that men failed to invent cotton pickers although they had the knowledge to do so. The case for cotton may be different. No effective machine appeared until after the tractor with its power takeoff. A cotton harvester was patented as early as 1850. Pneumatic suction machines were tried, especially after that method had proved effective in ginning. Strippers, which pulled the boll from the stalk, had been tried by 1871.

Cotton was grown in places other than the Old South. From 1865 onward, the center of production slowly shifted West. California had become an important producer by 1877, when inventors tried to make a cotton harvester there. The success of the combine in the wheat fields of California may have stimulated the effort.

Cotton presented several problems which none of the early machines handled effectively. The bolls matured and opened unevenly, and too many leaves were picked along with the cotton. Varieties that matured more uniformly were bred. These had more open pods so that a well-designed machine could get the cotton without the trash. The cotton picker also had to have a continuous source of power which could pick the cotton in a stationary position. Steam power proved too expensive and awkward for adaptation to cotton harvesting.

188

Currier and Ives print, 1883. Courtesy Library of Congress

The threshing and winnowing of grain, which could be performed in the field, had no real counterpart in cotton processing. Ginning effectively removed the seeds, but not leaves and dirt. Trash adhered to the cotton lint and mangled into the staples in ginning, which caused problems for spinners and weavers. Textile manufacturers needed clean cotton, and only handpicking produced such cotton. A shortage of human pickers led some cotton growers to "sled" their cotton, that is, to pull strippers across the field. This happened as early as 1914 or earlier. The practice was uneconomical and was done solely to prevent the total loss of the crop. As of 1914, the major problems of suitable types of cotton and a source of sustained power were unsolved.

HARVESTING TOBACCO

As late as 1971, no effective tobacco harvester had been invented. The onetime slave-labor nature of the crop handling tended to confuse the issue. Tobacco, like cotton, seemed to require human labor for most efficient harvesting of the crop. Slavery seemed to offer the best method technologically. Tobacco picking remained a form of stoop labor whether the harvesters cut the leaves from the stalk or cut the whole plant and removed the leaves later. The quality of the leaves varied so much that human judgment was needed in the picking and sorting. Unlike flax or hemp, tobacco had no strong durable fibers. The success of the corn harvester suggests a similar machine might have been developed for tobacco. The selecting and curing process would not have changed much with the application of a machine to the harvest. Suckering and cutting by hand remained typical of tobacco harvesting well into the 20th century.

INVENTION OF THE GRAIN REAPER AND BINDER

Inventors, manufacturers, and farmers undertook two concurrent lines of improvement in the simple grain reaper between 1865 and 1914. One sought to reduce the labor used in harvesting and speed the work by mechanical bundling and tying of the sheaves with a binder. The other approach was to reduce the overall time of reaping and threshing by doing both in the field with a combine simultaneously. The combine appeared first, but the binder was more successful initially. However, the combine was destined to be the more successful machine with the application of power from internal combustion engines.

Wire binders were invented first, but the expensive wire caught in the machinery and defied disposal after the harvest. Wire entered threshing machines, mixed in the straw, and killed the unlucky animals that ingested it. John Appleby invented and patented the twine binder in 1874. William Deering licensed Appleby's binder and added it to Deering reapers. Soon the twine binder was widespread. Other manufacturers used the Appleby invention on license with their own binders. Deering combined the Marsh style of harvester with Appleby's binder and produced 3,000 machines by 1880. Twine had completely replaced wire in self-binders by 1882, and in that year Cyrus McCormick produced and sold at least 15,000 of the new machines.

The market was glutted with consequent reductions in price. This accelerated the rate at which farmers shifted to the binder. In the wheat areas of the Midwest the binder soon appeared everywhere. On their smaller acreages, eastern farmers frequently had their reaping done on a custom basis.

The number of reaping and harvesting machines manufactured rose from 60,000 in 1880 to 250,000 by 1885. In the wheat areas, especially the Far West and the Midwest, wheat growers brought in most of their crop with machines. The twine binder rapidly replaced older self-rake machines. Lower costs meant greater profits and less need for harvest labor. The savings in time and labor enabled an increase in wheat acreage and in the total crop.

The combined reaper-thresher continued in use, especially on the West Coast. The harvester, a reaper which allowed men to ride while binding the grain, was popular through the 1870s. The machine suddenly declined in use in the 1880s after the self-binder hit the market. As many as 100,000 harvesters were in the field in 1879; hardly any were left except as curiosities by 1890.

In the 19th century, judging the added efficiency became a matter of confused statistics and learned speculation. High and low estimates were possible for the advantages of the binder over either reapers or harvesters. The binder saved two men over the harvester and five men over the reaper. This was significant in figuring the wage bill, but not much use in figuring production increases. The self-binder could have a wider cutter bar and thereby cut a larger swath than the harvester. The work performed by a three-horse binder varied from 10 to 16 acres a day.

IMPROVEMENTS OF MOWERS

The mower outlasted the harvester or self-binder. The self-raking reaper persisted in many places for cutting flax or harvesting crops other than wheat. Equipment manufacturers eventually combined mowers with the self-raking apparatus so farmers could use the machines for a variety of crops. The mower cut the green crop, and the reaper cut the matured crop. Forage crops had to be cut sharply, but cereal stems could be partly snapped. These crop characteristics determined the mower design.

The cutter bar of the mower had to move rapidly to cut the comparatively thick stands of grass. The mower sickles moved back and forth about three times for every one time of the reaper sickles. Put another way, the mower made about 22 cuts every 3 feet, compared to about 7.5 cuts for the same distance for the reaper. Forward speed was essential to develop enough power for effective cutting. Slow operation resulted in a heavy draft, which strained the animals, and either jammed the sickles or jumped over clumps without cutting. The cutter bars had to be sharper and better designed than those of the reaper. The width of the cut had to be narrow for a faster operation with a minimum of trouble. The 18-foot cutter bar of the giant combines was out of the question. Mowers ranged from 5 to 6 feet in width.

The cut hay was left in the field to cure. Whether stored as hay or silage, the forage had to be dried, gathered, and stored in separate operations.

ADVANCES IN THRESHING AND WINNOWING

Mechanical advances were made in threshing and winnowing. After threshing, the grain was winnowed separately as late as 1872. However, by that time, more threshers had effective winnowing fans. Winnowing was carried out by a fan blowing the grain through screens. The forced air dislodged chaff, weed seeds, and small particles of stems and leaves from the grain. This refuse was deposited with the straw, usually by wind current. Winnowing devices were standard by 1880. Hand winnowers continued in use, mostly by frugal farmers with small acreages.

Each machine performed three operations. A revolving cylinder (developed before 1860) knocked the grain free of the straw. Various belt or apron systems carried the straw away, elevating it, and dumping it onto a haystack. The labor of at least two men in stacking straw was saved. Automatic sacking devices were also added to the threshers. By the 1880s, the moving apron and elevator for disposing of the straw had become commonplace in all wheat country. The typical threshing machine had a capacity of 750 to 800 bushels a day.

Two types of steam engines were widely used. The most popular, the portable steam engine, had been in use since the 1850s. Pulled into the field by horses, this engine ran the thresher off a drive wheel. These engines could also power saws and run some dairying equipment. The owners could keep the machines profitably in the Midwest and the Far West. Like the expensive thresher, the steam engine was also used for custom work. The second type of steam power was the less common steam tractor which usually had a wheel for power take-off by means of a belt. These rigs were too cumbersome to be used in extensive custom work. They were usually part of the equipment of a large wheat farm.

Between 1800 and 1914, the capacities of the threshers were at least doubled, and automatic twine cutters, feeders, weighers, and blowers were added. The new machines required even more powerful steam engines which were usually pulled into the field. Custom harvesting became the rule nearly everywhere.

By 1900, custom harvesters generally started with the harvest in Texas in June and worked their way north with the seasons to the Dakotas and Montana, ending in September or later. Itinerant workers, called hoboes, were essential to the harvest. Railroad detectives

Poster, mounted and unmounted horsepowers, *ca.* 1880.

often ignored the hoboes as they traveled the freights in search of work. The hoboes did not become unpopular on the railroads until the combine made them unnecessary for the harvest.

Horse-powered threshing was still done in places like New York, Pennsylvania, and Virginia. The comparatively small farms and the smaller amounts of grain made the farmer-owned threshers and sweep powers practical. Treadmill power persisted in use in New England and out-of-the-way places where farmers threshed in small volume, chiefly for family use. In the Ohio and upper Mississippi valleys, the four- to ten-horse sweep powers continued in use through the 1870s, though steadily declining in importance. Manufacturers continued to advertise sweep powers well after 1914.

On large wheat farms, the steam engine and the custom rig dominated by 1890. Around 20 percent of the wheat crop was threshed by horsepower in 1880. By 1914, an even smaller proportion was threshed this way. The gasoline engine, stationary or tractor, began to replace the steam engine around 1900, and by the start of World War I the internal combustion engine had made considerable inroads on the steam threshing rigs. Simultaneously, the combines began to replace threshing rigs.

DEVELOPMENT OF THE COMBINE

Changes in the threshing rig, including the self-feed, elevator, and other automatic devices which operated off horsepower, led to further modifications. Farmers put the thresher on wheels and ran its mechanism from a ground wheel. Farmers preferred the header, a reaper which cut the grain close to the head and therefore handled less straw than other reapers. Then the grain was fed directly from the reaper or header to the following thresher. The next step involved combining all of this machinery into one machine. Without patent information, few official records document the transformation from several implements to one. In 1878, wheat growers still used three separate horse-drawn implements, including the sacking wagon, in some dry land areas, notably the San Joaquin valley. About 1880, the true combine began to appear in experimental versions.

An older tradition developed with the practice of using several horse- or mule-drawn machines in the field. A prototype combine, introduced in California in 1854, had performed well, but it burned in 1856. It was copied and improved. In the 1870s, changes took place in threshing equipment, and the new threshing equipment and the early combine converged to create an improved combine.

The combine needed an immense amount of power. Economy of

scale required large scale, not only in the fields, but in the combine it-self. The cutter bars were around 18 feet wide (regardless of the type), and about 20 horses or mules in two ranks of ten pulled it. One ground wheel commonly ran the cutter bar, and the other powered the thresh-ing machinery. The driver sat on an elevated platform high above the machine and the teams. Four men customarily rode on the combine to sack the grain and throw the sacks onto wagons alongside. Between 1884 and 1890, machines with 30-foot cutter bars and weighing fifteen tons had appeared. They required teams of forty or more animals. Steam tractors were soon used.

Four major problems were solved by about 1900. The combine could not be used on hilly ground because it would topple over; on rough ground the long cutter bar caught; runaway horses often stripped the gears; and the threshing machinery worked off the un-even forward speed of the combine. In the 1860s and 1870s, inventors worked out methods for raising and lowering the wheels on one side or another. By the 1880s, sidehill and up-and-downhill levelers solved the problem of toppling on all but the hilliest land. Raising and lowering the cutter bar was solved before the combine came into wide use. In 1886, Henry Holt figured out a link chain method of trans-mitting power from the wheels of the machinery to prevent the stripping of gears. If the team ran away, the chains broke, thus saving the mechanism. The harvest crew replaced the links in the field with a minimum of downtime. Smooth, effective operation of the threshing and elevator mechanisms came with the application of an independent power source. Before 1900, small steam engines powered this portion of the machinery independently. After 1900, the stationary gasoline engine replaced the steam engine.

The combine sharply increased productivity per man because it reduced the threshing crew and cut a larger swath. Twenty horses and four men with an 18-foot cutter bar could harvest at least 36 acres a day by 1880. Production rose as the size and efficiency of the machines increased. Smaller machines could harvest 23 to 29 acres a day under normal field conditions. As many as 45 acres could be harvested in a sparser stand.

Combines with internal combustion engines for working the threshing equipment, which animals still pulled, were first used widely in Argentina. The early gasoline tractors could not pull the giant combines. The gasoline engine combine came into wide use in the United States after World War I. The internal combustion combine could still operate efficiently at a lower weight and with a narrower swath.

Steam combines of the 1890s and early 1900s required too many men to fire and maintain them to give any real economy of labor. Potentially they could cut as much as 90 acres a day, but they usually

Harvester, 18 horses, Washington state, *ca.* 1888.

cut only about 60 acres under normal field conditions. In addition, operators always faced the danger of fires in the field set by sparks from the steam engine.

Steam tractors had a short, romanticized period of glory in pulling the combines, but the gasoline tractor replaced them after 1912. The majority of farmers harvested by horsepower before 1914. On the other hand, the application of steam and gasoline already promised important changes in harvesting even before World War I.

HARVESTING AND STORING FODDER

For stockmen, preparing and storing fodder and ensilage paralleled the grain grower's threshing. Between 1865 and 1914, metal replaced wood in most of the fodder implements, and the endless apron was developed as a method of loading silos and stacking straw and hay. These devices were modifications of inventions worked out for grain harvesting.

The fodder had to be cut and deposited in regular swaths to make later curing and gathering easier. Mowers had achieved this goal by the 1850s. Next the fodder had to be dried, for it spoiled and also heated internally if wet or damp. Spontaneous combustion could occur. The grass dried in the field after being raked into windrows. Machines called tedders were invented to fluff up the cut grass in the field. Europeans, probably the British, seem to have invented tedders around 1800, but these were not widely used in the United States until after the widespread use of mowers. In mowing the horses tended to press the hay down.

Apparently tedders of American design came onto the market shortly after the Civil War. The metal tedders had prongs or forks extending near the ground, and reciprocating mechanisms, including cams, kicked the hay along as the tedder passed over the hay. The power came from ground wheels. The forks were spaced about a foot apart, set to kick at varying intervals from one another, and a tedder usually carried about eight forks of two or three tines each.

By the 1870s, farmers used two general types of mechanism for dumping the accumulated hay. In one form, the farmer pulled a lever which raised and tipped the tines of the rakes and left a pile of hay in the windrow. In another design, the operator threw a gear and a camlike mechanism into gear, and as the rake wheels moved forward, the tines rose by power from the wheels and the load dumped. All types of rakes were invented before 1865, but they came into widespread use only in the 1870s.

Most of these rakes were sulky rakes. The old roll-over wooden rake did well enough for a man on foot. The object of the riding spring-toothed rake was to get the man off the ground. Greater speeds could be achieved by the sulky rake. Rakes were usually between 8 and 12 feet wide. After 1900, manufacturers improved their rakes by using more high-grade steel in wearing parts and by putting the gears into boxes.

The old rakes formed elongated piles which the pickups could not handle readily. In the 1890s, side-delivery sulky rakes appeared which used two methods to push the hay to one side in a rounded pile. One rake had tedder forks set to kick to the side instead of backward. The other rake had a set of wheel rakes also set to move around at right angles to the movement of the machine and push the hay to one side of the rake. The wheel rakes made a fairly round pile. Their side delivery mechanisms also fluffed the hay in the process of moving it to one side, lessening the need for the tedder. By 1910, the side delivery rake was the preferred rake, especially where mechanical hay loaders were used. By 1914, side delivery rakes predominated, except on the Great Plains. Plainsmen did not use much hay, and the dry climate made the use of tedders superfluous.

After the farmer had the hay in windrows, dried and ready for storage, he had to remove it from the field. Manufacturers developed field pickups and loaders, and farmers soon used them along with the side delivery rake. In common with other farm implements, patents for pickups and loaders abounded before anyone actually manufactured a successful machine. The first successful hay pickup and loader appeared around 1866. The hay had to be picked up off the ground, carried to the wagon, and distributed in the wagon. The machine had to get all of the hay, operate off a ground wheel (which meant it worked only while moving), carry the hay without dropping much or shaking leaves off the hay (particularly clover). The raking and delivery had to be synchronized so the machine stacked as rapidly as it picked up the hay.

The first loaders had their forks on the rear of the wagon, and a belt or apron moved the hay toward the front of the wagon. Later versions used cylinders with tines to pick up the hay and move it. Power came from ground wheels connected to gears, belts, and endless chains. These mechanical marvels cost too much as a rule to warrant the investment. On the market in the 1870s, they became popular only in the 1890s.

The increased interest in hay pickups and loaders may have been related to the development of the silo and ensilage. The volume of fodder needed for silage encouraged the development of powered cutters with traveling aprons to put silage into the silo. The first of these cutter-loaders appeared in 1876, only three years after the initial experiments with the silo in the United States. This first cutter was designed to handle corn as well as hay.

As early as 1853, H. L. Emery patented a horse-powered, screw-operated hay press. Others soon came onto the market. These machines worked slowly enough so that only farmers with large and efficient operations used them. After the baler made the bale, the press had to be returned to its original position to receive the next load of hay. In 1872, P. K. Dederick of Albany, New York, brought out his "Perpetual Press." His press provided for continual action along a horizontal rather than a vertical plane and much speeded the pressing of hay. Many improvements followed and steam power was successfully applied about 1882. Bales of hay took up less room in the barn, and they could be more easily moved than loose hay. A big market existed for baled hay. The farmers usually stacked baled hay and covered it with a tarpaulin. From 1865 to 1914, most hay was stored out-of-doors in a haystack. As of 1914, no good way had been found for shipping silage. Fodder could be baled and moved easily in large volume. The continuous hay baler and the silo had arrived within a year of one another.

In this period just preceding the advent of the internal combustion tractor, farming required large numbers of horses, mules, and oxen. The great use of animal power was facilitated by improvements in growing, harvesting, and processing animal feeds. Farm machinery did not simply lighten man's work load. The earliest machines allowed men to use the superior strength of animals without being hampered by their inferior intelligence. Machines made it possible for animals to do man's work with a minimum of supervision.

CHAPTER SEVENTEEN

THE TRACTOR APPEARS ✺ 1892–1914

WHEN MEN SOUGHT to harness animal power they strove to give the animal some human capabilities. The object of animal-powered inventions was to use the superior strength of the animal to do such tasks (reaping or churning, for example) as the animal could not otherwise do. True industrialization of agriculture—the substitution of other sources of power for animal strength—began in the late 19th century. The agent, symbol, and method were all one—the internal combustion tractor.

By 1900, some 5,000 steam tractors were made a year. The machines in use were trivial compared to the farmwork done. The steam tractor showed what nonanimal power could do, but it had serious shortcomings, too. The engine consumed large amounts of fuel and water and proved too difficult to fuel when in use. It also was very heavy and hard to move.

Because of the high steam pressures and the nature of the fuels, makers of steam tractors used heavy steel and mounted the engine on a steel frame. Improvements, which made them lighter and more efficient, occurred about the time the first gasoline tractors appeared. Early tractor engineering probably owed more to the steam tractor than to the somewhat earlier gasoline automobile. However, the tractor's importance arose from its automotive capabilities. Few businesses are operated across land surfaces on such a large scale as farming. Road building, mining, and transportation also move across large spaces to function.

Men developed traction from both steam and gasoline stationary engines. In Britain (and to a lesser extent in America), stationary steam engines were hooked to pulleys and cables to pull plows. The engines then had to be moved by some other power source each time the plow passed across the field. Some engines were even floated in canals and worked across nearby fields. Mining railways also used the stationary engine to draw cars. The San Francisco cable cars and skiing tows

Steam tractor, Springfield, Illinois, 1890.

continue to use the method. The farm tractor not only drew objects, but was mobile itself. It also could be a source of power for other machines by using a flywheel.

Automotive power offered several advantages over animal or human power. The internal combustion tractor surpassed animals in strength, even though the first machines were not very powerful. The tractor could provide power for other implements and machines, either standing still or moving. The amount of power produced did not depend on the forward movement of the tractor, as with the ground wheel of a machine pulled by an animal. The shifting from moving to stationary operation did not require complicated hitching and animal management. Large amounts of power might be supplied to a machine such as a harvester, while a small amount of power slowly moved the rig across the field.

Furthermore, tractors provided exceptional potentials for automation because of the vast amount of power which could be transferred to hydraulic and electric devices. Soon the tractor could drive plows into the ground forcefully at the beginning of a furrow and lift them out at its end. Ultimately, closed-circuit television would allow the operator to watch the field behind him. From the beginning, the tractor had potentialities which animal-driven devices never did.

Tractor, Hart Parr, 1903. (Smithsonian Instituti
Collections)

DEVELOPMENT OF THE GASOLINE TRACTOR

The automotive tractor began essentially as a stationary engine, connected to wheels and placed on the frame of a steam tractor. The first successful internal combustion engine was invented in Europe. The German Nicholas Otto patented a four-cycle internal combustion engine in 1876. He and others, notably Gottlieb Daimler, used the engine to power vehicles. Automobiles, rightly called horseless carriages, had appeared by 1888. A tractor with this kind of engine had

been built in the United States by 1889, but it developed too little power to operate a plow.

The first useful gasoline tractor appeared in the wheat fields of South Dakota in the harvest of 1892. John Froelich of Iowa put the machine together by mounting a stationary gasoline engine on a wooden and steel frame. Froelich used a simple two-cycle engine rather than the four-cycle engine which powered many automobiles. This meant that his engine had to have a flywheel which, once started, tended to stay in motion. Thus by inertia the wheel moved the piston through the full cycle. His one-cylinder engine had a hand pump on top so the operator could pump up pressure in the cylinder for each successive firing. It did not move very fast and tired its operator.

Mounted on a heavy frame, the entire combination weighed about 9,000 pounds and produced around 30 horsepower. It still weighed less than a comparable steam engine which developed only 14 horsepower and usually weighed around 12,000 pounds. The first internal combustion tractor was huge and unwieldy. But it worked without breaking down altogether and required far fewer workers to keep it in the field because no one needed to supply fuel constantly. It carried its own fuel supply and needed no attendant water tanks or fuel carts.

From 1892 onward, inventors sought to increase the drawbar power, lessen the engine weight, and reduce the size of the wagon frame on which the engine was mounted. They also tried to achieve more dependable operation and more available power. Another advantage of the gasoline tractor, aside from labor savings, was that it did not have to get up a head of steam. However, the first tractors were not easy to start either.

EARLY MANUFACTURERS OF TRACTORS

From 1892 on, some tractors were made, but not on an industrial scale. By 1902, several farm machinery companies were selling gasoline tractors. These were specially built by the machinery companies. In 1901, Charles Hart and Charles Parr of Iowa made their first successful Hart-Parr tractor. (A tractor of theirs made in 1903 is at the Smithsonian Institution and still runs.) They formed the first tractor company in Charles City, Iowa, in 1903. Stimultaneous developments took place in Europe, but the British tractor resembled an automobile. Hart-Parr mass-produced a regular line of tractors. The early Hart-Parr tractor weighed around 10 tons. Within a few years, machines of 30 horsepower which weighed only around 5 tons were made. In 1906, Hart and Parr gave the name "tractor" to their machines.

By 1909, the *Farm Implement News* mentioned thirty-one trac-

Tractor, Nichols-Shepard, 1912.

tor makers who produced some 2,000 tractors that year. As late as 1920, long after the gasoline tractor had shown its practical superiority, several companies built and sold some 2,000 steam tractors.

IMPROVEMENTS IN THE GASOLINE TRACTOR

The very first tractor had a gasoline fuel pump so the engine would not stall in the field. The fuel pump appeared later in automobiles. Self-starters came after 1914, when the storage battery was developed. The impulse coupling for the magneto and the use of a dry battery for the development of an initial current also appeared early. The tractors usually had large wheels to get the best ground power. The first successful Caterpillar tractor was patented in 1904 by Benjamin Holt, who also had made many improvements in grain combines. Commercial production of crawlers and tracks was under way by 1908. The track tractor inspired the development of the tank, introduced by the British during World War I.

The big breakthrough before the war was the development of the unit design, or frameless tractor. The tractor ceased to be an engine mounted on a frame and became a single, unified machine. The unit design greatly reduced the size and weight of the tractor.

The Wallace Cub, made in 1913, was the first frameless tractor. It had three wheels—one in front and two behind—to make it easier to use in row crops. The Cub was not the first three wheeler. The Bear (of what was later the Massey-Harris line) was the first three wheeler, built in 1912. Significant too was the introduction of tractors with more than one cylinder. The added power assured more reliable operation and ended the need for the fly wheel. International Harvester introduced the two-cylinder tractor in 1910. However, most tractors had one-cylinder engines up to World War I.

The mass production of tractors and the standardization involved in the process made it possible for farmers to buy replacement parts without having them specially machined. Enclosed transmissions, antifriction bearings, and parts made of steel alloys made tractors more useful for general farm work. Many of the advances came after 1914.

Figures for the number of tractors made and for the number on farms indicate their impact on American farming. From possibly 5 gasoline tractors in 1900, the number had risen to at least 17,000 fourteen years later. Around 30 corporations made the machines at that time. Although comparatively expensive, tractors soon declined in price because of mass production techniques that producers effectively applied during the war.

Before World War I, tractors did not have power takeoff. By 1912, the giant combines were usually pulled by gasoline tractors, by then far superior to their steam competitors. But the harvesting machinery continued to be powered by small gasoline engines mounted on the combine. This method of harvesting persisted through the war. The power takeoff was not essential in these large machines. It benefitted the smaller operator, especially the corn grower.

Tractors had flywheels of various sizes, which acted as a power-takeoff mechanism. The early tractors needed the centrifugal force and inertia of the wheels to carry the pistons through the cycles of compression and firing, especially when the tractor had only one piston. Later tractors had the wheels to use in belt hookups, although the wheel was no longer needed to keep the engine going.

USES OF TRACTORS

The tractor was useful for a variety of tasks. Farmers used tractors on jobs for which they would almost never have used the steam engine because of the excessive expense and time needed to put steam into use. The gasoline engine started promptly, worked efficiently, and

Caterpillar, Holt. First track tractor ever used in agricultural work, 1914.

was easily stopped. It could be put where the work was, rather than having the work brought to it. The tractor could run washing machines and other useful machines on the farm or in the home. The farm corn grinding mills for preparing feed, power saws, and similar machinery could be powered from the belt. Such machinery had long been adapted to belt power, so no great modification was needed. In addition, traveling aprons and other loading equipment, such as augers (especially for silos), could be powered from the tractor. Formerly, they had been powered from stationary engines. Water could be pumped with tractors. With virtually no rural electrification, the tractor and its belt power played a versatile role on the farm.

The tractor not only lightened labor in the obvious way, but it required less maintenance than animals. No studies were made of what the farmer did with the saved time, but since the census showed continuous increases in average farm acreages between 1890 and 1920, part of the saved time must have gone into farming more land. Also, since the need for animal feeds declined, land once used to raise horse fodder could be shifted over to human food crops. So, without adding a single acre to farmland, the amount of land used for commercial food production gradually began to increase.

By 1914, the tide was already turning in farm population. Farmers, declining in relative strength in the population since 1880, began to decline absolutely sometime between 1910 and 1920. The movement of farmers off the land had begun, and the change owed something to the introduction of the tractor. The effects of these great and pronounced changes became most apparent during the two great world wars.

CHAPTER EIGHTEEN

TOTAL WARS: WORLD WAR I, WORLD WAR II ⚔ 1914–1945

No SET OF EVENTS made such an impression on western civilization as the two world wars of the 20th century. Europe first felt the heaviest impact in 1914–1918; then America suffered the transforming trauma in 1939–1945. Terror, destruction, and killing caused fundamental changes in every aspect of life. Heretofore, no age or society had had the technical competence to organize and prosecute killing on such a vast and pervasive scale. Gigantic armies and navies were put into action. These forces reduced manpower in industry and agriculture which resulted in some shortages of production. To keep the armies intact, nations had to channel all production to support otherwise unproductive men, who nevertheless engaged in exhausting work.

Simple wage, profit, and price controls could not effectively channel all goods where needed. Certain classes of commodities were in truly critical undersupply. To meet the problem of real shortages, governments then had to resort to coercive distribution or redistribution, generally by rationing. Governments assigned economic goods in certain quantities to certain consumers regardless of any natural economic flow. Most nations, then, developed two currencies. Money, the first currency, bought the products. The second currency gave the buyer the right to purchase the commodity in question. In most cases the second currency took the form of ration stamps. Even so, in no case did the permission currency guarantee that the buyer could get the commodity. Rather the buyer received permission to purchase, provided he could find the goods.

Both total wars taught the great lesson that a government could mobilize its entire economy. Total manpower mobilization was possible largely because machines could be substituted for men on farms and in factories. The methods worked out in war could be applied on a smaller scale to national goals in peacetime. Not every method of mobilization would meet with popular approval, and rationing was abandoned as soon as practicable in most places. But monetary, fiscal, and other controls, including subsidies on a large scale, could be used to encourage individuals to pursue goals which accorded with those of the entire society. The wars were important not only in their own right, but also in the repercussions they caused in the periods which intervened between wars. Land policy and tenure, marketing of farm products, and science and technology on the farm will be viewed within this context. Broad developments emerge with an examination of statistical changes between 1910 and 1920 (including World War I) and 1940 and 1950 (including World War II). The changes reveal that war brought increased farm production, no matter how disagreeable the process.

FARM TENANCY

Tenancy tends to reflect prosperity or depression in the 20th century. Generally, the more numerous the tenants, the less prosperous the farm community, and vice versa. In 1910, tenant farms came to 37.0 percent of all farms, but they fell to 32.9 percent in 1920. The depression in the 1930s increased tenancy, and 38.7 percent of all farms in 1940 were tenant farms. This fell even more sharply during World War II, and in 1950, tenants made up only 26.8 percent of all farmers.

The census distinguished farm population as such only after 1920. In 1910, the total rural population was 49,973,000; this rose to 51,552,000 in 1920. In 1940, farm population alone only came to 30,216,000 and fell to 23,077,000 by 1950. But as the number of farmers fell, the size of farms increased as shown in Table 18.1. The total amount of land in farms rose from 878,798,000 acres of farmland in 1910 to 1,159,789,000 acres in 1930.

Economists, sociologists, and historians often speak of "war years-abnormality" and of "normal years." Normal years would seem to have

TABLE 18.1. Average Size of Farms, 1910–50

Year	Size
	acres
World War I	
1910	138.13
1920	148.24
World War II	
1940	173.99
1945	199.96
1950	215.49

SOURCE: *Historical Statistics*, 1957.

neither wars nor depressions. However, war and depression years far outnumber all other years in the 20th century so wars must also be considered as normal!

STOCK GRAZING HOMESTEAD ACT OF 1916

During both world wars, legislation facilitated westward expansion. The last important area for migration was the Great Plains. The population of the ten Plains states increased from 5,365,149 in 1910 to 6,031,968 in 1920. The Stock Grazing Homestead Act of 1916 was designed to transfer public domain to responsible stockmen and farmers on a far greater scale than before. Congress wanted to get the government out of the business of trying to control overgrazing and soil erosion. The act became law on December 29, 1916, when the United States was on the verge of entering the European war.

The Stock Grazing Homestead Act allowed farmers or preferably stockmen to get 640 acres of "non-irrigable, non-timbered land chiefly valuable for grazing and raising forage crops. . . ." (39 *Stat.* 862). The secretary of the interior was to designate the land to be homesteaded. He did not in fact specify much land although stockmen had applied for some 20 million acres by 1917. Most of the land was used for wheat instead of grazing. Congress bypassed a reluctant secretary of the interior and opened all public lands in New Mexico, Colorado, Utah, and South Dakota, but the land could not be entered until 1918.

The act became the most popular homestead legislation ever passed. Between 1916 and 1945, some 93,080,980 acres were homesteaded under various acts. Of these claims, farmers perfected 32,712,176 acres under the Stock Grazing Homestead Act. The act so cut up the public domain that its value for grazing or farming was greatly reduced although a large amount still remained under federal control.

Between 1914 and 1919, the land in wheat increased by 27 million acres, with winter wheat accounting for 21 million acres. Of this, more than half was grown in the northern and central Plains. Private sales of land, mostly by ranchers, added to the wheat lands, particularly in the northern Plains. Even before the Stock Grazing Homestead Act was implemented, many settlers began moving onto the northern and southern Plains. Between 1909 and 1919, the Plains counties of Montana, North Dakota, and South Dakota (taken as a whole) increased their harvested wheat acreage from 2,563,000 to 4,903,000, and at the same time the number of cattle fell from 2,278,000 to 2,170,000.

EFFECTS OF WORLD WAR I ON AGRICULTURE

The United States's land policy on the eve of its entry into World War I made vast acreages available for bread grain production. The warning of impending involvement and the time to prepare made the course of events in the farm sector rather different from the situation pertaining if the country had been at war from the beginning. World War I increased the market and also created higher prices. In response, American farmers produced heavily and no doubt contributed to the Allied victory. The war brought an increased general prosperity to the United States, shared by industry, commerce, and farming as shown in Table 18.2.

TABLE 18.2. Commodities in Thousands and Yearly Average Prices, 1914–18

	Year	No.	Av. Price
		head	
Cattle	1914	59,461	$38.97
	1915	63,849	40.67
	1916	67,438	40.10
	1917	70,979	43.34
	1918	73,040	50.01
Hogs	1914	52,853	10.51
	1915	56,600	9.95
	1916	60,596	8.48
	1917	57,578	11.82
	1918	62,931	19.69
		bu	
Corn	1914	2,523,750	.708
	1915	2,829,044	.676
	1916	2,425,206	1.137
	1917	2,908,242	1.456
	1918	2,441,249	1.520
Wheat	1914	897,487	.975
	1915	1,008,637	.961
	1916	637,572	1.434
	1917	619,790	2.047
	1918	904,130	2.050
		lb	
Cheese	1914	367,000	.146
	1915	440,000	.142
	1916	422,000	.175
	1917	472,000	.225
	1918	415,227	.290
Butter	1914	1,684,749	.298
	1915	1,750,613	.298
	1916	1,793,113	.340
	1917	1,644,029	.427
	1918	904,130	2.050

SOURCE: *Historical Statistics*, 1957.

One estimate of the total profits of wheat farmers showed an increase from $56,713,000 in 1913 to $642,837,000 by 1917! Net cheese exports have shown excess exports only three times in the 20th century, twice during war or near war. In the years 1914–1919, the United States exported 14,787,000 pounds of cheese more than it imported. The boom in nonexport items was also important to the farmer because full employment and higher wages increased the domestic urban market, too.

GOVERNMENT CONTROLS AFFECTING AGRICULTURE

The government encouraged farmers to produce in the national interest by appealing to their patriotism and by manipulation of the market. When the United States entered the war on the Allied side, President Woodrow Wilson immediately accelerated food production. Congress passed a Food Production Act on August 10, 1917. The law authorized the secretary of agriculture to make food surveys, encourage production of needed foods, and help farmers in securing labor. Congress also passed the Food and Fuel Control Act on August 10, 1917, and set up a food and a fuel administration to regulate both industries. The act forbade hoarding or monopolizing of supplies and sought to secure fair distribution of limited commodities. The president was empowered to fix a minimum price per bushel for wheat and directly control the distribution of food, fuel, feed, fertilizer, and farm machinery.

President Wilson appointed Herbert C. Hoover as Food Administrator and Harry A. Garfield as Fuel Administrator. Hoover and the Food Administration had wider powers than generally recognized. The United States did not resort to rationing; instead, citizens were encouraged to observe wheatless days, meatless days, and sugarless days. Urbanites were urged to have gardens, later called "victory gardens."

Efforts were made to regulate the marketing of farm products. On August 18, 1914, Congress passed the Cotton Futures Act, which contributed some stability to the market. On August 11, 1916, the United States Warehouse Act and the Grain Standards Act were passed. Both these acts attempted to prevent fraud and cheating on the part of processors and merchants.

Through the market machinery the Food Administration influenced prices and production. First of all, the government gave direct subsidies. The Food Administration set up a Grain Corporation which had money and authority to buy grain on the open market. It bought enough wheat to keep the price around $2.20 per bushel. Some other grains and even meat were purchased in order to support prices. Furthermore, Allied purchasing officers had to go through the Food Administration which set prices for the benefit of farmers and sometimes of urbanites. For example, the administration established minimum prices for swine of $15.50 per hundredweight. However, hog raisers resented the pricing policies for foreign purchasers, the U.S. Army, and the prices set for corn for feeders. They claimed they could have received more from the Europeans in a free market. The price setting for foreign purchasers and buying licenses kept a fairly good control on supplies, distribution, and prices. The Food Administration claimed to have helped wheat farmers achieve profits of about $643 million in 1917 and $815 million in 1918.

Farmers not only benefitted from direct action aimed at them, but they felt the impact of indirect action as well. For example, on December 28, 1917, the Federal Government seized the nation's railroads by executive order and put them under the secretary of the treasury. In the process of making a nearly defunct rail system work again, the Railroad Administration (with the help of the Department of Agriculture) established certain days and seasons for the shipment of livestock to terminal markets. The two agencies thus regulated not only the flow of the product, but restricted the alternatives of sellers and buyers alike. Rail operation by the government did keep transportation charges within bounds for the farmers.

As early as September 1916, Congress created a Shipping Board, authorized it to build or buy ships, and appropriated $50 million to do this. By 1918, the United States, largely through the Emergency Fleet Corporation, had spent about $3 billion and had doubled the U.S. shipping tonnage. Lower transport charges and, therefore, lower costs for the total commodity, resulted from subsidies to the merchant marine.

FEDERAL ROAD BUILDING

In both world wars the most significant technological changes occurred in the greater use of tractors and automobiles. The truck particularly changed the marketing procedures and possibilities for the American farmer. But the change would hardly have been as dramatic without the intervention of the Federal Government in road building. A movement for federal aid to states for road building had begun in the days of the bicycle craze. Few hard-surfaced roads had been built, however, and federal aid had been negligible.

The Rural Post Roads Act passed in 1916. Although ostensibly for speeding mail delivery, the subsidy of $75 million for the first three years was put under the control of the secretary of agriculture. The Federal Government agreed to pay half the cost of improving state roads up to a maximum of $10,000 a mile. In 1915, the United States had 276,000 miles of hard-surfaced rural roads, and this had risen to 369,000 miles by 1920. The program of federal aid began with little opposition, largely because it had been hooked to the war effort. However, federal road building could be almost infinitely expanded to the great benefit of farmers, truckers, and others.

WORLD WAR II AND U.S. AGRICULTURE

The intervention of the Federal Government on a large scale in agriculture during World War I set precedents for the next war. World War II saw an alteration and increase in governmental farm activity. In spite of marked changes in law and the economy, the New Deal, however, always had to battle vociferous and sometimes vicious opposition. In contrast, World War II gave the government an umbrella of patriotism for the new programs. Necessity forced the government to take action which it had previously avoided because of the hostility of the farm community. The earlier U.S. experience probably had some influence on American decisions. The government did not take over the railroads during World War II.

Almost at once the European war had impact on American commercial farming as exports fell from 30 to 40 percent below the average of the ten preceding depression years. Grain exports, for example, fell 30 percent between September 1939 and September 1940. The fall of Norway, Denmark, the Netherlands, Belgium, and France had removed them from American trade. However, the market improved dramatically as the Germans continued their successful invasions.

The United States had unofficially entered the war on the side of the Allies in 1940 when U.S. destroyers began to hunt German U-boats. The navy also convoyed materials to Britain. Congress and the president anticipated more American participation in the war with an accompanying regulation of the national economy.

FOOD FOR THE ALLIES

Congress passed the Lend-Lease Act in 1940 which gave food, among other things, to the Allies. The program amounted to an export program for farmers. Lend-Lease also stimulated U.S. industry and increased employment. The Commodity Credit Corporation sold its holdings to Lend-Lease and other agencies. The Food Distribution Administration bought foodstuffs on the U.S. market and sent them to Europe.

Agricultural policy, as in the earlier war, aimed at increasing the production of certain products. The depression had caused a shortage of capital equipment—tools, implements, and farming machinery. The government could direct and increase commodity production simply by entering the market and buying commodities at guaranteed prices. Hogs, poultry, eggs, and some dairy products and vegetables were the first priorities for production.

In 1940, an Agricultural Division was set up under the National Defense Advisory Council with Chester Davis as administrator. His division was to maintain parity of farm products with other product prices and to prevent shortages of farm products. The Agricultural Division soon was transferred to the Office of Agricultural Defense Relations in the Department of Agriculture.

Congress also stimulated the production of certain so-called basic commodities through price supports and a special act of May 26, 1941. Prices of cotton and cottonseed products rose by 75 percent and of cereal grains by 50 percent. Cotton and cereal production both shot up. By the end of 1941, farm income was higher than at any time since 1929 although direct cash payments accounted for only 13 percent of farm income.

In May 1941, Great Britain and the United States formed the Anglo-American Food Committee. About the same time, the Office of Agricultural Defense Relations was moved to the Department of Agriculture. The two federal agencies joined in planning aid to Britain and indirectly set the stage for the agricultural preparedness of the United States for its participation in World War II.

Although the British had tried to build up supplies of bread grains, meat, and tobacco, they soon had shortages of milk, butter, eggs, cheese, fish, and fats. Beginning in early 1941, the U.S. Department of Agriculture and its agencies had encouraged the production of these products. The difficulties in transporting these highly perishable commodities were met by direct government subsidies to processors to find ways of dehydrating eggs and milk.

The Federal Government responded to the war need by entering the market and buying products away from the domestic purchaser—a method which had worked fairly well during World War I. Early in 1941, the British asked for 22 million cases of canned milk annually. In response, several federal agencies bought 40 million cases of canned milk between May 1941 and July 1942. Milk came to be in short supply in the United States, and in December 1942, the Food Distribution Administration put 2 million cases of milk on the domestic market to relieve the shortage.

UNITED STATES DECLARES WAR

On December 7, 1941, the United States officially declared war. The armed forces began to purchase food in large quantities for the ever-growing services. The army and navy had begun purchasing even before direct U.S. involvement in the war, and other agencies had also begun stockpiling agricultural supplies. Now the armed forces stockpiled on a huge scale and took most of the food which was diverted from civilian consumption. As in World War I, military purchases also were used to keep prices high for farm products.

In 1942, Americans sent some 23 percent of their cheese and dried skim milk, 7.2 percent of their butter (to the Soviet Union), 10 percent of their eggs (dehydrated), and smaller percentages of other foods to the Allies. During the first half of 1943, some 45 million pounds of beef and 8.1 million pounds of butter went to the Allies. Secretary of Agriculture Claude Wickard reported to Congress that: "The British in 1942 received about 20 percent of their edible fats from us, about 10 percent of their meat (largely pork), and about 20 percent of their other protein foods such as milk, cheese, and eggs." The Russians also received large shipments of foods, particularly sugar and fats.

Counting Lend-Lease aid, other aid, and military purchases, the Federal Government became the chief food buyer in the United States. By 1943, it requisitioned and paid for 30 percent of all butter produced, 80 percent of the lower grades of beef, 40 percent of all other beef, 30 percent of the veal, 35 percent of the lamb and mutton, and 50 percent of the canned fruits and vegetables. The American farmer could not push production high enough, and shortages resulted in rationing as early as 1942. The prices paid to farmers rose in spite of price controls at the retail-consumer and wholesale levels.

In 1942, the first of the clearly wartime laws was passed, which raised price supports to 90 percent of parity for certain needed commodities. In addition, Congress reduced the penalties for going over the quotas if the crop were needed. Simultaneously, Congress increased the list of supported commodities. Production rose in response although the number of farmers continued to fall.

On June 5, 1942, the administration set up a Food Requirements Committee under the War Production Board. Then Roosevelt and Churchill established a Combined-Food Board to make best use of the food production of the Allies. American farmers found themselves controlled, regulated, and exhorted by the Department of Agriculture to increase production. At the same time, farmers found themselves under the control and regulation of a welter of other agencies.

Labor problems fell within the scope of the War Manpower Commission, problems of securing farm machinery went to the War Production Board, and the movement of products came under the

Office of Defense Transportation. The Department of Agriculture sought to raise or maintain farm income in relation to the rest of the economy, while the Office of Price Administration tried to hold prices down for consumers. The War Production Board had to balance tanks and planes against tractors and milking machines, and it did not always rule to the advantage of farmers.

On December 6, 1942, the president appointed Claude Wickard, secretary of agriculture, to serve also as War Food Administrator. In March 1943, Chester Davis took over the job until June when Marvin Jones became administrator and served out the war. These men occupied roughly the same position as Herbert C. Hoover during World War I. The new War Food Administration, however, had greater powers of coercion.

Necessity and previous experience combined to make the system work—farmers were effectively mobilized and resources were redirected. Many farm laborers joined the armed services or became industrial workers. Those who remained on the farms had to be used efficiently. Farm machinery, fertilizers, insecticides, herbicides, and other capital goods were in short supply. Transportation of products was hampered by shortages of gasoline, rubber, and vehicles. Chemicals, from fertilizers through insecticides, were produced in competition with munitions. All these shortages occurred while demands for farm production by both the military and civilian sectors continued to increase.

The experiment stations and extension services were effectively mobilized to teach new, improved farming methods. New seeds, fertilizer, machinery, and animals were developed with government assistance and guidance. The U.S. farmer responded by raising 50 percent more food annually during World War II than during World War I. This tremendous increase was accomplished on less land and with probably 10 percent fewer workers.

The organization of the nation for total war during World War II resulted in far-reaching changes in farming. The regulations were largely an extension and intensification of those of World War I. Both wars left a residue of experience that could serve also in peacetime.

Most importantly, agricultural products had become a government cartel by the end of World War II. Even commerce was actually controlled and usually financed by the federal treasury. The domestic consumer, thanks to wartime prosperity, clamored for more and better food and he had the money to buy it.

The system of cartelization allowed the nation to prosecute the war effort successfully. At the same time, it brought a measure of economic gain to farmers with a minimum of loss to urban consumers. On the whole, while Americans did not want rationing again, they seemed to have accepted continuing government economic intervention. The cartel system of government-sponsored monopolistic marketing for agriculture would be retained for a long time.

CHAPTER NINETEEN

THE PEOPLE AND THE LAND ⚓ 1914–1945

P EOPLE live on the land and the land holds people. Those who ignore the land come finally to ecological distress. The problems of the man-to-land ratio became apparent in the United States only during the 20th century. Some prophets of the 19th century had peered into the future and found the prospects alarming. But the problems of population and land use appeared comparatively late in American history. When they suddenly did appear they took an odd and unique turn.

The percentage of farm population had been declining since the 1880s, although the number of people on farms actually increased until World War I. Then the farm population began to fall absolutely. The introduction of the tractor and other new farm machinery had made it possible for fewer people to handle more land. The size of farms continued to grow, but no corresponding increase in the number of people was necessary. Wartime opportunities for urban employment absorbed any surplus farm population.

Not that the shift from the land to the cities took place smoothly across the years. The depression of the thirties apparently drove many back to the farms. Authorities surmise this reverse migration began in 1936 or 1937. Rural farm population went from 31,393,000 in 1920 down to 30,158,000 in 1930, and to 30,216,000 in 1940. By 1950, farm population had fallen to 23,077,000 and fell regularly thereafter.

In 1914, farm units numbered about 6,447,000 and rose irregularly until 1935 to about 6,814,000 farm units in the United States. By 1940, the number declined to 6,350,000; it had fallen to 5,967,000 by 1945. Farmland increased from 909,627,000 acres in 1914 to 1,141,614,000 acres in 1945. The average size of a farm was 137.9 acres in 1914, 142.8 in 1925, 154.7 in 1935, and 191.3 in 1945.

Immigration was cut back sharply by the United States' entry in World War I and the immigration laws of 1922 and 1924. However, population stood at 99,118,000 total in 1914 and at 123,188,000 by

Picking cotton, Louisiana, 1934. Courtesy Library of Congress

1930. The Great Depression slowed the population growth of the United States which had increased to only 132,594,000 people in 1940. Within the next five years, almost 8 million were added. From 41,999,000 in urban areas in 1910, cities had grown to 74,424,000 in 1940; and the rural nonfarm population had risen from 20,159,000 in 1920 to 38,693,000 in 1950. The wartime baby boom continued through the 1950s. The urban population alone of 1950 was 88,927,000 —just 10 million short of the total population of 1914. The total U.S. population divided by its land area gives its population density. Tables 19.1 and 19.2 give a comparison of the United States with the United Kingdom and Belgium.

The density of farm population in relation to farmland fell constantly. It was 21 per square mile in 1920; 19.5 per square mile in 1930; and 18.2 per square mile in 1940. The migration to cities accelerated from 1914 to 1945. In this period, immigrants from abroad declined in numbers and southern blacks began a steady movement into the cities of the nation. Here they crowded into the ghettos which had once sheltered other bumpkin immigrants who had since moved on. It became fashionable to insist that the economic stagnation of the immigrants from the South resulted solely from racial discrimination. After all, Irish peasants had been absorbed quickly and successfully into the cities, but the black was still an outcast after centuries. The argument ignored the fact that no matter how long the blacks had been in the Republic, they had only recently stopped being unsophisticated farmers.

If more land went into farms and the farms got larger, where did the land come from? Some of it was federal land, some state-owned

218

TABLE 19.1. Comparative Population Density*

Year	Density per Square Mile, United States	Density per Square Mile, Britain	Density per Square Mile, Belgium
ca. 1910–11	31.0	446.6	600.8
ca. 1920	35.6	467.3	636.9
ca. 1930	41.2	488.6	698.3
ca. 1940	44.2	509.5	712.2

* Figured on land areas of 2,977,128 square miles for the United States, 94,209 square miles for Great Britain, and 11,775 square miles for Belgium.

land (school grants and such), and some was railroad land. Most of the statistics on this land are contradictory and confusing, but some parts of the land can be accounted for. Certain states never had federal land in any appreciable amounts within their boundaries. Land of the original thirteen, plus Vermont, Kentucky, Texas, and bits of other areas, had been regularly under state control.

TABLE 19.2. Population of the United States if as Dense as the United Kingdom and Belgium*

Year	If as Dense as the U.K.	If as Dense as Belgium
ca. 1910	1,329,585,364	1,788,658,502
ca. 1920	1,391,211,914	1,896,132,823
ca. 1930	1,454,624,740	2,078,928,482
ca. 1940	1,516,846,716	2,120,310,561

SOURCES: World Almanac, 1960; U.S. Census Reports, relevant years.
* United Kingdom population based on the difference between population in 1931 and 1951 since the United Kingdom held no census in 1941. The Belgian Figures for 1930 and 1940 are for census reports of 1933 and 1938.

CHANGES IN LAND USE

The vacant public domain of the United States, most of it western land, shrank significantly between 1914 and 1945—from 291 million acres in 1914 to 170 million acres in 1945. The largest transfer came between 1914 and 1929. More than 100 million acres went to farmers. Between 1914 and 1934, 125 million acres were taken out of the public domain by farmers.

At the same time, land was steadily being withdrawn from agricultural use. In 1911, Congress provided for the purchase of national forests from private owners and states in the East, and an act of 1924 allowed purchase in the West as well. Table 19.3 summarizes the

TABLE 19.3. Land Taken out of Agriculture, 1915–45*

Category of Land	Acres Withdrawn, 1915–45
Land purchased for national forests	15,532,000
Land put into national parks	16,140,000
Land taken up by primary and secondary roads	4,056,960
Land absorbed into cities	2,146,688
Total	37,875,648

SOURCES: U.S. Bureau of Public Roads; *Statistical Abstract,* relevant years; U.S. Department of Transportation, *Highway Statistics,* 1967.

* To figure the land taken by roads, take the average right-of-way of each class of road as given by the Bureau of Roads and multiply by the mileage of roads in each category to get total land surface. For land taken into cities, take the 1960 figures for the land area of cities over 100,000 population, and figure back proportionately to the same population density of 1910, and subtract estimated area of 1910 from area of 1945 to get land area removed. Increase the answer for land in cities by 25 percent to account for the cities under 100,000 population.

acres withdrawn. The land withdrawn represented about 3.3 percent of the total farmland in 1945. The withdrawal of farmland took place during an immense expansion in agricultural production that reflected the astounding effects of farm science and technology.

INDIAN LANDS AND GOVERNMENT POLICIES

In 1887, Congress decided the Indians, on their ever-shrinking reservations supposedly received in perpetuity, should be integrated into the mainstream of white American society. The Dawes Act provided for the dissolution of the reservations. The Indians were to receive parcels of land of their own selection: 160 acres for an adult with a family, 80 acres for unmarried adults, 40 acres for widows and orphans. The Indian received a trust certificate which he could not sell and which he held for 25 years, at the end of which time he could receive a fee simple title. Congress hoped thus to prevent the Indians from being cheated of their land after they had selected it.

Indian lands under the jurisdiction of the Bureau of Indian Affairs fell steadily: 155,632,312 acres in 1881; 78,372,185 acres in 1900; 71,646,796 acres in 1911; to 52,651,393 acres by 1933. The losses between 1881 and 1933 came to 102,980,919 acres, but it does not show how many acres had gone to Indians under trust certificates. However, trust acreage in 1949 came to only 16,534,060 acres.

In spite of the loss of their lands, Indian tribal loyalties remained strong. Few Indians became successful farmers. Generally they had land inferior from an agricultural viewpoint, since they had no guidance in selecting good farmland; and, more importantly, most of the land to be distributed was located in places where even the most skilled white farmers could not succeed with only 160 acres. The Great Depression, however, provided the impetus for the changes which reformers had long advocated.

THE INDIAN REORGANIZATION ACT
AND THE TAYLOR GRAZING ACT

In 1934, Congress passed the Indian Reorganization Act and the Taylor Grazing Act which greatly altered federal land policy. The first allowed the Indians to hold onto their reservations. Where the reservations had been fragmented, the law provided funds so that the Department of the Interior could purchase lands to give the Indians solid blocks of land. The second act allowed the president to withdraw parts of the public domain from either homesteading or sale. It also allowed the Department of the Interior to charge grazing fees on the land so withdrawn.

Although the new legislation allowed the tribes to reestablish themselves as formal entities, the law chiefly changed the nature of land tenure. The amount of land available to the tribes increased: 52,651,393 acres in 1934, 55,362,949 in 1945, and 56,004,670 in 1949. Most of the added land was purchased by the Department of the Interior from the farmland of white farmers. The Indians used much of the land for grazing so the shift amounted to a withdrawal of cropland.

The tribal lands were held in trust by the Federal Government. The Indians were not likely, without difficult legal steps, to be dispossessed, except for special bills depriving the Indians of specific land parcels. On balance, the outright expropriations, carried on for three and a half centuries, came to an end.

HOMESTEADING IN THE ARID WEST

Most of the theoretically unoccupied public domain was in Alaska and the western states from the Great Plains to the Pacific Coast. Farmers had homesteaded, purchased, or received the land from railroads and states and put it into crops. On the arid land of the West the biggest upsurge in homesteading took place between 1913 and 1925. This coincided with new discoveries in dry farming which made occupation of much of the area technically feasible. But dry farmers had moved onto the Plains in times of relatively heavy rainfall. Then

Daytime dust storm, Amarillo, Texas, Apr. 1936.
Courtesy Library of Congress

came several years of moderate drought. Dust storms carried off topsoil, seed, and plants and deposited the windblown garbage on other
acres. Farmers abandoned millions of acres, open to wind and sometimes to flood erosion as well. Tax delinquency and abandoned farms
increased, especially from Texas to Montana and the Dakotas.

The Federal Government reluctantly decided to halt further
homesteading and to attempt reclamation of the ruined, often abandoned acreages. On the most arid land, which even optimistic farmers
had avoided, stockmen grazed animals on a first-come, first-served
basis. They formed associations that had power but no legal standing
under federal law. Overgrazing also produced fearful erosion. The
government decided to charge for the privilege of grazing and thereby
limit further destruction of the land, which had accelerated during
the 20th century. All of these problems became dramatically evident
with the first of the great dust storms in the thirties.

TAYLOR GRAZING ACT AND
THE END OF THE OPEN RANGE

Something had to be done to conserve the land itself and Congress
passed the Taylor Grazing Act on June 28, 1934. The act authorized
the president to withdraw land from entry by homesteaders and pur-

chasers if the land was unsuitable for farming. The act also enabled the land users to join the grazing districts that were under federal supervision. Some 80 million acres were included initially in the grazing districts where the government could limit the number of animals grazed and charge for the use of the grass.

The act also gave the Department of the Interior power to develop water power, control soil erosion, and dispose of federal land not in the grazing districts. This land included about 86 million additional acres in 1934. On February 5, 1935, President Franklin D. Roosevelt issued an executive order, withdrawing the entire remaining 166 million acres of public domain from entry. In 1936, Congress amended the act to include 142 million acres in grazing districts. The Open Range came to an end.

The Forest Service had imposed grazing fees and successfully collected them from those who grazed animals in the national forests. Cattlemen and sheepmen resisted vigorously; but they found that violent resistance counted for nothing, and resistance in the courts invariably failed. They therefore had to take political action. In 1925, congressional hearings on the Stanfield Bill were held across the West. This bill would have given permanent rights to grazers in the national forests. Livestockmen wanted perpetual tenure and they wanted it free or nearly so. In Congress, however, they represented a minority interest. Both the departments of Interior and Agriculture opposed the Stanfield Bill which failed to pass.

Other similar measures failed because the big fight of 1925 had represented a full trial of strength. The defeat of the measure marked a turning point in the attitude of westerners toward grazing and the public domain. The stockmen determined that the public domain would never be turned over to the Department of Agriculture and its Forest Service. As it happened, turning the land over to the Department of the Interior in 1934 did not improve the tenure position of the old-time cowboys.

INCREASING FARM PRODUCTIVITY

Paradoxically, as agricultural development seemed to be curtailed, farm production increased. As some farmers were ruined, others became prosperous; as available land contracted, exploitation of resources increased. In fact, they managed to raise more even though cropland decreased, as shown in Table 19.4.

The common claim is made that the yield per acre rose little or not at all. Science and technology thus had little impact on total production except that science and technology let fewer people farm

TABLE 19.4. Production Increases, 1915–44

Year	Tractors on Farms	Wheat	Pork
		bu	lb
1915	25,000	1,008,637,000	13,935,217,000
1919	158,000	945,403,000	13,985,843,000
1929	827,000	800,649,000	15,581,878,000
1939	1,445,000	708,582,000	17,078,849,000
1944	2,354,000	1,032,660,000	20,503,755,000

SOURCE: USDA, ERS, *Statistical Bulletin 233*, 1966.

more land. But the facts show steadily increasing gross production in spite of occasional ups and downs. Land programs and land tenure had very little effect on total production.

In the case of productivity per acre, for example, any national or regional figures conceal real changes in capabilities. If highly productive land is taken out of production in wheat, and land which had previously been unfarmable is put into wheat, the yield per acre may fall. An acre which yielded nothing before, when made capable of yielding even 5 bushels, is an acre which has increased in usefulness. This, in fact, happened with wheat on the Great Plains.

Wheat continued to advance onto the Great Plains, especially the northern Plains, through World War I and into the 1920s. Acres harvested in the Plains counties of Montana, North Dakota, and South Dakota rose: 2,563,000 acres in 1909, 4,903,000 acres in 1919, and 6,646,000 acres in 1924.

TENURE CHANGES ON FARMS

Tenure changes may have had some impact on the application of science and technology. The larger the farms and the fewer the farmers, the greater the application of advances in technology and science. The fewer the tenants, the more innovative the agriculture.

Table 19.5 shows the trend in tenure changes. Apparently all

TABLE 19.5. Tenure Changes, 1910–50

Year	All Farms	Full Owners	Part Owners	Managers	Cash Tenants	Other Tenants
1910	6,361,502	3,354,897	593,825	58,104	712,294	1,642,382
1920	7,448,343	3,366,510	558,580	68,449	480,009	1,974,795
1930	6,288,648	2,911,644	656,750	55,889	489,210	2,175,155
1940	6,096,799	3,084,138	615,750	36,351	514,438	1,846,833
1950	5,382,162	3,089,583	824,923	23,527	212,790	1,231,339

SOURCE: *Statistical Abstract*, 1960.

categories went down, but not at the same rate. Only the number of part owners increased between 1910 and 1950. Part owners rose 38.9 percent when every other tenure category was declining. Though the number of managers fell only 59.5 percent through the years, few were left by 1950.

The number of farms decreased only 15.3 percent between 1910 and 1950. The number of full owners fell only 7.9 percent; the percentage of full owners was greater in 1950 than in 1910. Full owners had come to dominate U.S. farm life. Other tenants, chiefly sharecroppers, declined 25 percent. Cash tenants dropped a spectacular 70.1 percent while the number of farms fell only 15.3 percent. Some must have left the business altogether, and others became part owners or full owners. The percentage of the labor force engaged in agriculture fell from 27 percent in 1920 to 17.6 percent in 1940. The nature of land tenure had changed dramatically. The conditions of the survivors improved as the failures left. For the nation as a whole, the changes seemed beneficial.

CHAPTER TWENTY

CHANGES IN TRANSPORT, PROCESSING, AND SELLING ⚓ 1914–1945

H ISTORIANS and economists generally have asserted that the development of American agriculture resulted in created surpluses. However, the surpluses may have been centered only in certain commodities related to the ability of consumers to purchase. The decline in personal income in this period undoubtedly resulted in some surpluses because the ability of consumers to purchase determined the relation of demand to available supplies.

Between 1910 and 1945, the most conspicuous decline in personal income affected the bottom 30 percent of the population. The top of the bottom 30 percent showed improvements in wartime, but the bottom 20 percent showed relative improvement only in 1934 in the depths of the Great Depression. The bottom 10 percent of all economic groups had 3.4 percent of the total personal income in 1910 and only 1.0 percent in 1945. Gabriel Kolko has reviewed these developments in *Wealth and Power in America.*

One other group saw a steady diminishing of personal income before taxes: those in the top 10 percent. In 1910, they had 33.9 percent of all personal income; in 1945, only 29.0 percent. Most of the income increase fell to those in the second to the seventh tenths of the income groups. The upper middle class and the middle middle increased most. This group made up 60 percent of the population and controlled food prices. The middle group, in effect, determined the domestic consumption of farm products.

Not all income is from wages, but 90 percent of the people derive most income from wages, either in industry or on the farm. Tables 20.1 and 20.2 show some national averages. Not until 1941 did the average reach the level of 1929. Between 1929 and 1933, the gross national product fell regularly, and from 1934 to 1944, a steady rise set in, followed by a small drop in 1945. The 1929 level was

TABLE 20.1. Average Annual Earnings
in All Industries, 1914–34

1914	$ 627	1939	$1,264
1917	830	1941	1,443
1919	1,201	1942	1,709
1924	1,303	1943	1,951
1929	1,405	1944	2,108
1934	1,091		

SOURCE: *Historical Statistics,* 1957.

reached only after 1940, thus exhibiting the same trends which wages
showed.

CHANGES IN FOOD CONSUMPTION

Between 1914 and 1945, the per capita consumption of meat rose
slightly; dairy products substantially; and fruits, potatoes, and cereals
fell, sometimes noticeably. Oddly, potatoes and flour fell most sharply
during depressions and showed little response to war. Meat consumption responded to war significantly, and dairy product consumption
dropped markedly during both major wars. Fruit consumption fell
in wartime, rose in depressions, and slumped at other times. Table
20.3 shows total per capita consumption per year, including meat,
fruits, potatoes, flour, and dairy products.

After World War I, Europeans cut down on imports of U.S.
food. During the war 50 percent of British beef came from the United
States, but only 5 percent by 1923. Even cotton faced foreign competition, and the American market failed to take up the slack. Agricultural prices fluctuated with the prosperity of urban workers, espe-
cilly that middle 60 percent. In addition, eating patterns changed
with different work needs and improved knowledge about nutrition.

RAILROAD TRANSPORT CHANGES

In carrying freight, the railroads continued to dwarf all other types
of transportation. Railroads carried more than half of the ton mileage moved through the period. They accounted for 65 percent of the
ton miles transported. Very short hauls and local traffic do not enter
the figures. Table 20.5 shows the number and tonnages of freight
cars (1914–1945). The ton miles of freight carried followed closely the
economic conditions of the country. Well into the postwar period

TABLE 20.2. Per Capita Gross National Product

1929	$857	1937	$ 846	1945	$1,293
1933	590	1941	1,040		

SOURCE: *Historical Statistics,* 1957.

TABLE 20.3. Consumption of Major Foods, 1914–19

Per Capita (lbs per year)					
1914	1,414.8	1924	1,428.4	1939	1,390.1
1917	1,333.5	1929	1,421.3	1941	1,379.6
1919	1,340.9	1934	1,369.0	1945	1,358.2

SOURCE: *Historical Statistics,* 1957.

TABLE 20.4. Exports, 1914–44

	Percentage of Total Production					
Year	Wheat (& flour)	Rye (& flour)	Rice	Corn	Tobacco	Cotton
1914	37.4	30.9	12	2.5	38.1	54.0
1919	23.4	52.8	41	0.9	49.7	58.8
1924	30.4	86.0	12	1.0	37.9	60.5
1929	17.1	7.3	26	0.6	44.2	47.5
1934	2.0	...	11	0.2	36.4	52.3
1939	6.1	1.9	18	2.2	18.7	55.0
1944	13.6	7.8	33	0.8	17.6	15.6

SOURCE: USDA, *Foreign Agricultural Trade, Statistical Handbook, Statistical Bulletin 179,* 1956

TABLE 20.5. Freight Cars in Use, 1914–45

Year	Number	Average Capacity	Total Capacity
		tons	
1914	2,349,734	39.1	91,874,599
1918	2,397,943	41.6	99,754,428
1924	2,411,627	44.3	106,835,076
1929	2,323,683	46.3	107,586,523
1934	1,973,247	48.0	94,716,856
1939	1,680,519	49.7	83,321,794
1942	1,773,735	50.5	89,573,618
1945	1,787,073	51.1	91,319,430

SOURCE: *Historical Statistics,* 1945.

(after 1945), the railroads remained dominant in transportation. In 1914, the railroads moved 288.6 billion ton miles of freight. This rose to 450.2 billion in 1929 and fell to 270.3 billion in 1934. In 1945, the railroads moved 684.1 billion ton miles of freight.

USE OF DIESEL POWER

In the 1920s, the railroad began to substitute diesel power for steam, resulting in substantial fuel savings and faster service. In the 1930s, the diesels became more efficient, and the railroads also made greater use of electric power. Often the diesels generated electricity that actually moved the engines. Such engines could shift to electric power from power lines where available. The electric motor allowed faster acceleration, reduced the use of brakes on downgrades, and lessened the expense of repairs to rolling stock. The number of diesel locomotives rose from 11 in 1926 to 4,301 in 1945; steam declined from 66,847 engines in 1926 to 41,018 in 1945. Electric engines fell steadily for the period. The increasing ton miles of freight carried was accomplished by faster turnaround of freight cars, most of which was not related to railroad operation but to efficiencies in unloading and processing.

EFFECTS OF AUTOMOBILES

As the number of farm automobiles rose, the number of farms and farmers decreased, especially after 1935. In 1910, American farmers owned about 50,000 automobiles. In 1920, they owned 2,146,000, and the figure had climbed to 4,148,000 by 1945. Simultaneously the number of horses and mules declined with the increasing use of tractors. The urban horse population declined from 3 million in 1910 to 1.7 million in 1920. Thereafter the census listed urban and rural horses together, so few were the the horses in cities and towns. The sudden increase of automobiles, farm and urban, no doubt owed something to advances in automotive engineering. Closed cars finally outnumbered open ones in 1923. All-weather cars were of more importance to farmers than to urbanites. Self-starters, four-wheel brakes, safety glass, and more powerful engines came during the 1920s.

TRUCK TRANSPORT

Farm trucks were used chiefly to carry produce to markets. The number of such trucks rose from 25,000 in 1915 to 1,490,000 in 1945. The size and capacity of the farm trucks are not known, nor the size of the nonfarm trucks. The farm trucks were probably smaller, but they rapidly became the chief carriers of farm products.

The truck became important most rapidly in the dairy industry. Creameries and cheese factories, as well as whole milk dairies, had to provide pickup service or a competitor would. The farmer ceased bringing the milk to the railroad station to be carted off to the city on the "milk train." The greatest shift to trucks in dairying took place during World War II. The shortage of tires during the war helped speed the change. By 1948, 97 percent of the milk moved by truck and 3 percent by a combination of truck and rail. The following figures show that in 1948 farmers shipped 91 percent of their grain by direct truck haul, some 83 percent of fruits, 82 percent of vegetables, and 88 percent of livestock. Other commodities were even more affected by the truck. Farmers shipped 96 percent of their cotton, poultry, and eggs and 99 percent of their tobacco by truck in 1948.*

Trucks began to be important for livestock transport during the 1920s. The shorter transit time cut losses. By 1932, cattlemen sent more than half of their livestock to market in trucks. The savings did not all go to the carriers or the Teamsters Union. The savings mostly went to the producers, who thus increased the profits of their operations. As in most technological advances, economies favored the big producers.

ROAD BUILDING

Widespread use of the truck depended to a large extent on building hard-surfaced roads, which began expanding rapidly in the 1920s. From 1921 to the beginnings of the Great Depression, Congress appropriated 75 million dollars annually to road building. Between 1919 and 1929, the road-building efficiency of work crews rose from 4½ to 18 running feet a day. The miles of surfaced rural roads shot up from 521,000 in 1925 to 1,721,000 in 1945. Although still primitive by later standards, they allowed trucks to replace the railroads

---

* Margaret Purcell, *Statistical Findings of Survey of Transportation from Farms to Initial Markets* (Washington: U.S. Department of Agriculture, Bureau of Agricultural Economics, Mimeograph, 1949).

as the prime carrier of farm products. The development of the pneumatic tire also contributed to the change. The low capital requirements for trucks made it easy for carriers to enter the business, and the flexibility of service was important. The truck could go from the farm to the market without transshipment. Trucks could give fast service for small loads and for perishables such as vegetables and dairy products. Farmers with trucks could carry for others when they no longer needed the trucks for their own produce.

U.S. MERCHANT FLEET

Efficiency in transportation and handling encouraged farmers to produce more. Their products could be put at the ship dock cheaper than before. If the U.S. merchant marine had been as effective as in the days of the clipper, the United States might have dominated the world markets. Apparently the United States could retain a merchant fleet only with federal subsidies. About every imaginable subsidy was given the shipbuilding industry except cash after World War I. No appreciable increase in shipping resulted. Tonnage fell steadily during the 1920s and 1930s, and America was again undersupplied when World War II began. A War Shipping Board successfully increased the mechant fleet after 1941. By 1945, the merchant fleet came to 30 million tons, about ten times the amount of 1941, but the government began to divest itself of its holdings in 1946. The additional wartime tonnage did assure the movement of farm products to the world market.

FOOD PROCESSING DEVELOPMENTS—QUICK-FREEZING

As transportation changed and became more efficient, so did the related industry of processing and handling. Monumental changes in food processing occurred in the years 1914–1945. Food preservation aimed at making products survive rough handling and long trips. Other benefits followed more or less accidentally. Metal cans added weight, but they unquestionably made handling, transport, and storage possible for perishable commodities. Canning factories substituted machines for many of the old hand processes of the industry. Eventually, processors could can anything that could stand cooking.

The most significant change in food processing and handling occurred in the quick-freezing method of preservation. Inventors experimented with frozen foods in 1914 and developed fairly successful methods in the 1930s. Clarence Birdseye, inventor of the plate-type freezer, sold his process to General Foods. In 1932, General Foods began selling frozen products through its subsidiary, Frosted Foods. Competitors appeared, but Frosted Foods (later called Birdseye) had half the business in frozen foods as late as 1938. Frozen foods (poultry, fish, meat, vegetables, and others) rose from 10 million pounds in 1934 to about 250 million pounds in 1938.

Many frozen foods, especially fruits and some vegetables, could be retailed in convenient packages. The foods were closer to the quality of fresh food than canned foods. Freezing, storage, and transportation facilities had to be developed, and the consumer had to have a freezer to store the food. During the Great Depression, inventors made some advances; but as trucks and warehouses increased during World War II, frozen food began to move out of the luxury category.

NEW MECHANICAL HANDLING METHODS

In the late 1930s (just before World War II), new mechanical handling methods came into use for warehouses. These included pallets, skids, and fork-lift tractors. Sometimes tractors or trucks moved products within plants. Belt conveyor lines had long been used, but movable belt systems came during the 1940s. These allowed direct movement from storage areas to loading docks for trucks and railroads. Canned foods and later frozen foods could be moved in quantity, with savings to the consumer and the farmer. In the late 1930s, clamp arrangements on tractors were developed to move and stack cotton. The labor reductions were impressive; one worker with a lift truck could handle as much as nine workers by hand. Costs and charges both fell.

For grain, the great breakthrough in the 1840s had been the invention of the grain elevator and endless chain buckets. About a century later, air pressure moved both the flour and its residues. Bucket elevators still delivered the grain into steel bins, but thereafter the product was moved by air pressure. Inert gas stored with the flour prevented explosions and halted spoilage. Pallets and fork-lift tractors moved the final products. Most of these devices came into use before the end of World War II.

MEAT MARKETING AND PROCESSING

Marketing and processing of meat changed after World War I. The use of terminal meat markets, such as Chicago or Omaha, changed

232

with increasing union activity in the 1930s. Increasingly the market and the processing became decentralized. Auction markets, local or regional packers, and packing in the country near the source of supply all steadily replaced the central markets. The figures showed a small shift during World War II, possibly related to meat rationing. The shift accelerated after the war. The new system was unquestionably more efficient. In addition, the packers' inability to exploit urban labor led to efforts to find other methods of cutting costs.

RETAIL OUTLETS FOR FOOD

Retail outlets for food were provided by grocery stores. The cost saving of chain stores was furthered by the development of the supermarket which began with the Piggly Wiggly chain in the 1920s and rapidly spread during the Great Depression. However, supermarkets did not dominate the food retailing business. The supermarket called for nearly complete self-service with savings in labor costs and other costs as well. In 1937, the A&P began operating supermarkets and reduced the number of stores by two-thirds by the 1930s and 1940s. The large supermarket used advanced methods of handling the produce of farm and processor. More people were served by fewer store employees with greater savings to consumers and presumably better prices for farmers.

Taking everything into consideration—technological changes in transport, processing, and selling—evident savings had been achieved. The first to secure direct economic advantage were those who instituted the savings: processors, transporters, wholesalers, and retailers. The next affected were the consumers which increasingly meant urbanites, although farmers also fell in this class and benefitted as well. The farmers as producers also shared in savings in costs, although unevenly and uncertainly.

All the technological changes did not bring farm prosperity, and certainly not for the small farmer. The solution of the farm problem apparently did not lie in the direction of greater efficiency in the work done off the farm. Neither did increasing technological advances on the farm seem to help much. The basic cause of the whole difficulty was the economic condition of most Americans and of the rest of mankind. The story of this chapter ends as it began. Too many people had too little income. Science and technology appreciably increased the prosperity of Americans and thus the prosperity of U.S. farmers, but in the long not the short run.

The people of the United States and their Congress directed the course of technological change through legislation. Efforts to make technological and scientific advances socially useful took a simple form in farming. The Federal Government simply forced the farmers into a monopolistic cartel and took over the marketing of the products. Few Americans expected farm relief programs to have a very profound influence on technological changes in farming.

CHAPTER TWENTY-ONE

PRICE SUPPORTS AND PARITY ✳ 1914–1945

T
HE LOSS of foreign markets, 1919 to 1940, brought agrarian distress. Many politicians and economists called for the Federal Government to set up and manage a cartel that would dump agricultural products on the world markets. Any loss would be recovered from somebody else. This simplistic understanding of the problem resulted in the cartelization of American agriculture.

The most publicized of the suggested remedies after World War I was the McNary-Haugen Bill. It was presented to Congress in revised form several times from 1924 to 1929. Historians thought these attempts came to a dead end; however, they did lead directly to long-range legislation and should receive more than passing attention.

George N. Peek, president of the Moline Plow Company, and Hugh S. Johnson, his assistant, decided (just before the Moline Plow Company went bankrupt) that their survival depended on farm prosperity. Their plan would restore the export market which had existed during World War I. This plan was embodied in the McNary-Haugen Bill. A two-price system for agriculture was proposed: one price for domestic consumers (protected by a high tariff) and another price for foreign consumers.

Although details of the proposed program changed, some general principles remained constant. A governmental agency would be set up to buy any given surplus, and the price would be based on prewar averages. The price would thus be higher than the current world price. (The idea also marked the beginning of the concept of parity.) The government would sell the surpluses as needed at home or abroad at world prices. If sold abroad, the product would sell at a loss. The loss suffered by the government was to be covered by what was thereafter called an "equalization fee."

In the first plan offered, the first purchaser of the commodity paid the farmer the difference between the U.S. price and the world price. The first purchaser, however, paid the difference in scrip not in money.

At the end of the crop year, the processor would have to redeem the scrip at some portion of its face value. That is, the first purchaser would buy back the scrip from the farmer, but only at part of its true value. The amount paid of the face value was determined by the losses the processor incurred in selling the commodity.

The proposed law was manifestly complicated so the supporters included six basic commodities: wheat, cotton, wool, cattle, sheep, and swine. The proposal did not include tobacco, corn, or butter, and the producers of these commodities opposed the plan, especially in the South. Eventually the McNary-Haugen Bill won the support of nearly all farm organizations and of Secretary of Agriculture Henry C. Wallace. However, he died in 1924 and the bill lost an important supporter.

Congress considered the bill in 1924 and in 1926. Each time it failed to clear both houses of Congress, although it may have passed one. The bill ignored too many special interests. Even its most vigorous supporters had to agree that the bill had a distinctively midwestern bias. Advocates easily solved the problem by adding more crops to the basic list.

In 1927, Congress passed the McNary-Haugen Bill: 51 to 43 in the Senate and 214 to 178 in the House. Herbert C. Hoover, then secretary of commerce, opposed the bill. He had been Food Administrator during World War I. Hoover probably influenced President Calvin Coolidge to veto the bill.

OBJECTIONS TO THE McNARY-HAUGEN BILL

Coolidge observed that the bill was designed to help certain farmers in certain sections at the expense of farmers in other sections. In retrospect this seemed to be a sound observation. He also feared the act would encourage one-crop agriculture. It was also a price-fixing measure, repugnant to the American free enterprise system. He felt the plan would be difficult if not impossible to administer. This last objection surely reflected an unfamiliarity with technological capabilities available to bureaucrats. Anticipating later Supreme Court decisions, Coolidge asserted that the equalization fee was not a true tax and was unconstitutional because it fell arbitrarily on a selected group of citizens. He correctly asserted that the act also called for an unconstitutional delegation of powers.

By this time the farm community had organized a farm lobby to push the bill into law over the objections of the president. In 1928, the farm lobby, now called the Farm Bloc, brought the bill up again and pushed it through with even more substantial majorities.

Once again Coolidge vetoed it, partly at the insistance of Herbert Hoover. The Farm Bloc did not have the votes to override the veto. Coolidge's veto message pointed out that the bill would only encourage overproduction with ever-greater surpluses. Subsequent experience indicated it would have done just that. The president noted that dumping our farm surpluses on foreign markets might well be resented.

EXPORT DEBENTURE PLAN

The Grange presented an export debenture plan that formed part of the background for subsequent events and legislation. The Grange's export debenture plan proposed to pay export bounties of 50 percent of the tariff rates on certain commodities. But the government was to pay in debentures, not in cash. These debentures could be used to pay tariff duties. If the business receiving the debentures did not import anything, it could sell the debentures to someone who did. The tariff thus paid the bounties. The debenture plan provided for the creation of a Federal Farm Board which would reduce or abolish debentures in proportion to increased crop yields. Thus aid would diminish as the need for it increased. Congress and many farmers felt the debenture plan was too indirect to help. They objected both to its aid plan and to the suggested coercion to be applied by the proposed Farm Board. Proponents presented the debenture bill to Congress several times, but Congress always rejected it.

AGRICULTURAL MARKETING ACT AND
THE FEDERAL FARM BOARD

Farm relief became a significant issue in the campaign of 1928 as farm prices continued below the cost of production. The Republicans nominated Herbert C. Hoover, the arch foe of direct help to farmers. When elected, Hoover called a special session of Congress to handle tariff and agricultural problems. The Agricultural Marketing Act, passed during this special session, clearly embodied Hoover's idea of farm relief. The act was intended chiefly to stimulate voluntary cooperation. Hoover believed farmers could in this way control production as big business did.

The act created a Federal Farm Board of eight members. The board, organized promptly with Alexander Legge of International Harvester as chairman, received a fund of $500 million to lend to cooperatives and to set up stabilization corporations. In first year the board loaned $165 million and unquestionably contributed greatly to the growth of cooperatives and to the improvement of crop marketing. The board also offered good advice on production restriction. Stabilization corporations had existed during the war under Hoover's direction. They could buy, store, and sell surpluses in specified commodities. The idea was to end seasonal marketing troubles, but the plan could not adjust to long-range price difficulties, especially if the price of stored commodities consistently fell.

Meanwhile, the U.S. Department of Agriculture and the land grant agricultural colleges almost defeated the objectives of the act unintentionally. These research institutions were always finding new ways to produce more on less land with fewer animals at less cost. This perfectly admirable effort wrought considerable achievements. But the research programs did conflict with ideas about crop and commodity restriction. The less was needed the more was grown.

GRAIN AND COTTON STABILIZATION CORPORATIONS

Meanwhile, cotton and wheat farmers needed something more than advice. As the great cash crops were grown in the North and the South, some political support was assured. Prices in both commodities continued downward. Cooperative marketing did not really work in the commodities or did not solve any price problems. So, in 1930, the Farm Board established a Grain Stabilization Corporation and a Cotton Stabilization Corporation. Copied after earlier corporations developed during World War I, they could buy in the open market and thus raise prices. They bought and prices rose.

The corporations were soon swamped with wheat and cotton, and prices never rose enough to allow disposal of their holdings. By June 1931, the Grain Corporation held 257 million bushels of wheat with no market in sight. The corporation stopped buying wheat. In July the price of wheat fell to 57 cents and dropped more later. The Cotton Corporation had accumulated 3,250,000 bales of cotton by 1932, and it stopped buying. The price of cotton promptly fell to 5 cents a pound. The Federal Farm Board went out of existence in 1933 with losses totaling $184 million, a large sum at the time.

AGRICULTURAL ADJUSTMENT ACT

When Franklin D. Roosevelt took office in 1933, his most pressing problem was the farm problem. His New Deal began as emergency legislation to save lives and farms. Long-term solutions could wait. The emergency laws took the form of the Agricultural Adjustment Act of 1933. The immediate aim was to reduce the acreage planted. By the time the act went into operation, however, much of the year's planting had already been done. As the men of the Agricultural Adjustment Administration put it, they "had entered a race with the sun."

In general, the New Dealers felt that overproduction was the main problem. Actually, the problem was underconsumption and could be traced directly to the wage levels of workers. In the 1920s, production of major crops had increased only 10 percent while the population had increased 16 percent. Per capita consumption of food in pounds fell 3 percent between 1929 and 1930. Consumption fell another 3 percent per capita between 1930 and 1933. By then millions of Americans were on the verge of starvation.

The Agricultural Adjustment Act of May 1933 gave the secretary of agriculture wide powers to raise price levels: by voluntary acreage reduction of certain crops, by direct payments to those who volunteered to cooperate, and by marketing agreements with processors and other handlers of farm products. He could license processors and associations of producers to eliminate unfair practices or charges. He could set up processing taxes if he felt them to be necessary. The taxes thus raised could be used to defray the costs of the program. And he could also use the money from the taxes to get rid of agricultural surpluses. In a way, Congress had passed the McNary-Haugen Bill. Most importantly, the Agricultural Adjustment Act carried with it the power to restrict production directly.

Voluntary contracts were drawn up between the farmers and the Department of Agriculture, or in legal fact the secretary of agriculture. For most of the crops the government rented the land from the farmer and did not plant anything. In effect the government paid the farmers not to produce. The covered, or supported, products included wheat, cotton, field corn, hogs, rice, tobacco, and milk. Congress intended that the farmer who rented his land to the secretary of agriculture should plant soil-conserving cover crops that would not glut the market. Thus erosion would be held down, and the price problem attacked at the same time.

Someone had to pay for all of this. Congress and the secretary of agriculture elected to pay the bill with a special tax on the initial processors: millers, meat packers, canners, or whomever. Surplus products of all kinds were to be stored on the farm. The farmer could use these crops as collateral for a loan from the government. The act covered only six crops. In 1934 and 1935, Congress expanded the list to include rye, flax, barley, grain sorghum, cattle, peanuts, sugar beets, sugarcane, and potatoes. The list eventually included 15 commodities.

NONRECOURSE LOANS AND PARITY

In the fall of 1933, the government began a program of nonrecourse loans for cotton and corn. The farmer borrowed money on his crop and put the corn or cotton up as collateral. At the end of the crop year he could sell the corn or cotton and repay the loan with interest if the price of the commodity rose. If not, he simply refused to pay the loan, and the government seized the crop. These loans were called nonrecourse loans because the government agreed in advance to take no other measures to collect the loans. On top of that, the government would willingly lend again the next year. The first loans were made at about 60 to 70 percent of parity.

Parity meant simply that the purchasing power of a certain unit of a commodity would be kept as it was during the period 1910–1914. Thus if a bushel of corn had purchased a pair of shoes (1910–1914), then a bushel of corn should bring enough in 1933 to buy a pair of shoes. Statisticians arrived at parity much as they later arrived at cost-of-living indexes. They made up an average "typical" shopping list. They also averaged the prices of commodities. Then statisticians put the two averages side by side which gave the parity for the base period. A similar mythical shopping list for the year currently in question would establish how much farm commodities would have to bring to give the farmer parity, using these averages.

If the market price of farm products did not equal parity, then the government made up the difference, chiefly through the nonrecourse loan method. The government never gave the farmer full parity, but only a percentage of parity. That is, if the farmer needed 10 cents more a bushel in 1933 to have the same buying power he had in the period 1910–1914, the government would give him only 6 cents more, or 60 percent of parity. Generally, the amount of parity he received depended on the amount of money Congress made available to the secretary of agriculture.

COMMODITY CREDIT CORPORATION AND AGRICULTURAL ADJUSTMENT ADMINISTRATION

The Commodity Credit Corporation (CCC) made the loans to the farmers. President Roosevelt set it up by executive order in October 1933. The government began to seize commodities in lieu of unpaid loans which then became the property of the Commodity Credit Corporation. Eventually the CCC became swamped in confiscated products.

The Agricultural Adjustment Administration (AAA) succeeded

Cricket fence barricades a road, Big Horn County, Montana, June 1939. Courtesy Library of Congress

at first. Eighty percent of the wheat growers took some of their land out of wheat; farmers cut their corn acreage by 50 percent and cotton by 75 percent. But the drought of 1934 made the act even more effective and prices almost doubled. Still, the contraction in acreage did not do the job. Some farmers lied, of course; small farmers particularly tended to maintain former levels of output. They were numerous and hard to check on. Farmers could not always be caught at the marketplace because they used wheat and corn on the farm. Restriction of acreage proved most successful in the cash crops of tobacco and cotton where the commodity had to go through some processing in order to be used.

In 1934, Congress put compulsory restrictions on cotton acreage, probably because they could be enforced. Between 1933 and 1935, the AAA reduced output of cotton by 10 to 13 million bales. For food crops, Congress began a policy of holding referendums on continuance of the programs. The food purchased was used in relief programs. In certain enumerated crops the secretary of agriculture could issue marketing orders. These came to essentially the old monopolistic or oligopolistic marketing cooperatives. Only now the government enforced the monopoly. These arrangements worked well, particularly in milk.

Between 1932 and 1937, net farm income rose 3.5 billion dollars in the United States. By 1935, farm income had risen 50 percent over what it had been in 1932. Of this rise, an estimated 25 percent took the form of direct payments to farmers. The rest of the increase resulted from acreage restrictions and reduced crops caused by drought. Even so, farmers did not get as much as they had in 1929 and did not in fact regain that level until 1941. Furthermore, 1929 had been a depression year for farmers, though by no means the worst depression year. As for direct payments, the Federal Government paid directly to farmers some 1.5 billion dollars between 1932 and 1936. The figure amounted to very little in terms of the whole economy, but amounted to a great deal for the farmers. For many of them the payments represented the margin for survival.

Three types of activity took place under the law of 1933 as finally amended. First, the amount of production was reduced. Second, the act put cash into the hands of the farmers. Third, the farmers and government joined in controlling the market and the prices received. The apparently successful act, however, was declared unconstitutional in 1936. The processors urged that the tax on them should be made a general tax instead. In the case of the *United States versus Butler,* the Supreme Court decided the processing tax was not imposed for the general welfare; therefore it could not be collected. Furthermore, the contracts between the farmers and the secretary of agriculture were declared unconstitutional. The Court held that only the states should regulate agriculture. The decision destroyed the act of 1933 and created an intolerable situation.

SOIL CONSERVATION AND THE DOMESTIC ALLOTMENT ACT

Congress immediately responded with the Soil Conservation and Domestic Allotment Act passed in February 1936. Congress rushed the act through so the checks could keep going to the farmers in spite of the Supreme Court decisions. As a stopgap law, it had some gaps, but it did fulfill its main object—to the give aid with unbroken continuity.

In effect, the new act declared that the formerly supported products now depleted the soil, and farmers should reduce their production of those commodities in order to conserve soil. The new act also set up the goal of income parity for each farmer instead of simple commodity price parity. The goal proved difficult to achieve. The act also provided direct cash inducements to farmers to use soil-conserving methods. The act received public support because of the devastating dust storms and the evidence of gully erosion.

Congress and the Supreme Court had voided all the old contracts between farmers and the secretary of agriculture. So instead of the contracts, the government merely invited farmers to cooperate. If a farmer did cooperate, he received cash payments. The payments averaged about $10 an acre for taking land out of production. In general, the soil-depleting crops happened also to be the crops found in surplus. Wheat, cotton, corn, tobacco, and sugar beets were officially declared to be soil depleting. The soil-conserving crops turned out to be those which did not contribute to the surplus. Payment came from the general funds without any special tax, and the law provided for an appropriation not to exceed $500 million.

The act proved as unsuccessful as a crop restricter as its predecessor. The acreage planted rose. Farmers also took their poorest land out of production and increased the yields on their better land. The continuing drought helped to keep production down, but sometimes production rose, especially in 1937. In addition, Congress did not always appropriate enough money to run the program properly. Still, the act apparently furthered soil conservation, and farm income rose until 1937. Income fell drastically for farmers in 1938, and stayed down until 1941 and World War II. But the recession of 1937 and the failures of the act seemed to call for new legislation.

AGRICULTURAL ADJUSTMENT ACT OF 1938

Congress passed a second Agricultural Adjustment Act in 1938. The act covered the five basic crops. Each year the Department of Agriculture had to estimate the necessary acreage for each crop. This allowable acreage was apportioned to the states and then to the counties, on the basis of past records of production. County farmer organizations then passed to the farmer his share of the county allotment, again on the basis of records of production. Each farmer was paid to stay within his allotments.

The act provided for three classes of payments. The first went to the farmer for keeping within the allotted acreage. The second payment went for carrying on certain conservation programs as detailed for states, localities, crops, and conditions. The third, parity payments, resulted from the demands of the farm lobby. The parity payments were designed partly to obtain an ever-normal granary with loans made by the Commodity Credit Corporation. All these payments could fluctuate between crops and crop years.

No farmer had to cooperate in any of this. If a farmer did not cooperate, however, then he did not get Class 1 or Class 3 payments. He could always get the conservation payments of Class 2 if he followed the rules. If he failed to cooperate fully, he got less preference in other governmental loans and programs. If surplus crops appeared, then the secretary of agriculture could invite farmers to undertake marketing restrictions. If two-thirds of the farmers in a crop voted to restrict production, then the Department of Agriculture could force compliance. The farmer, for example, would not be able to sell his grain. The Supreme Court upheld the constitutionality of the act.

Acreage restrictions and marketing quotas did reduce acreage in the years when they were used. Wheat acreage came to 81 million acres in 1937. This fell to 63 million in 1938. Then it went below 62 million acres and stayed below that amount until 1944. The same thing happened in all of the soil-depleting crops. Only acreage planted fell. In spite of all the efforts, production continued to rise because of increases in yields per acre in all crops.

Despite these efforts, prices declined about 25 percent between 1938 and 1940. At best the program merely kept things from getting worse. By 1939, direct cash payments for farmers, as distinct from indirect support, made up 35 percent of net cash farm income. This had fallen to 30 percent by 1940.

The programs failed to help the small farmers who made up most of the farmers. Payments, for example, did not go to tenants or to sharecroppers. Increasingly these people had to leave the land. As one southern landowner said, "I took the government payments, bought tractors, and got shut of my croppers." Landlords and farmers used payments from the government to buy fertilizers and better seed. Acreage taken out of corn went into soybeans. Farmers realized a double sort of return: one for conforming and one for selling the alternate crop. Small operators could not use these alternatives. Payments came to 709 million dollars in 1939. By World War II, the Commodity Credit Corporation owned 5 million bushels of wheat and about the same of corn.

The problem of surplus commodities remained unsolved and became worse. The larger, more efficient units also were the biggest producers. Land did not disappear as the small farmer left the business. Instead, total production increased. If surplus was the key to the problem, the federal program only contributed to the difficulty. By getting rid of the smallest and least efficient farm units, the program tended to increase both efficiency and surplus over the long run.

CHAPTER TWENTY-TWO

FARM MACHINERY TECHNOLOGY ⚹ 1914–1945

Technological advances became highly cumulative in all areas of life during the years 1914–1945. The total agricultural production rose spectacularly throughout the period as the proportion of people actually engaged in farming fell from 31 percent in 1910 to 17.6 percent in 1940. Yet the number of people fed each year by each farmer rose from 7.07 in 1910 to 10.69 in 1940, and by 1945, a farmer in the United States supplied 12.87 Americans and 1.68 foreigners.

Livestock husbandry, including dairying, needed 3,816 million man-hours between 1909 and 1913. This rose for the period 1932–1936 to an average of 5,159 million man-hours each year. Advances in animal husbandry had taken place in genetics, feeding, and veterinary medicine. Machines—such as milking machines, stall cleaners, and silage choppers—did not compensate for the larger numbers of animals. At the same time, the labor needs declined for crops such as corn, wheat, and cotton.

The extensive grazing on ranches and ranges had made slight demands on labor. As farmers and stockmen used grazing land for other farm enterprises and as farms became larger at the expense of grazing land, the stockmen had to refine their methods and use land more intensively. Less grazing land meant more feed crops were needed with more labor expended on those crops. Crowding animals increased chances of loss through disease and injury. These dangers called for more human supervision of the animals. The increasing use of dairy products also brought an increased need for labor. Bulk tank handling of milk hardly had begun during the 1940s.

Insofar as animal husbandry involved raising feed crops, the livestock raiser could benefit from advances in mechanical technology, as in the design of plows. Inventors had actually made many devices much earlier than they came into widespread use. Invented in the

244

19th century, the disk plow had been improved through the years. Roller bearings, notched disks—all made of high-grade steel—finally resulted in a first-rate plow for certain types of soil. The problem of sticky soils had never been quite solved by John Deere and others. As the disk revolved, sticky soils scraped off. The disk also rolled over stones instead of catching on them. It required more traction per square inch of soil surface turned than other plows. In the 1920s and 1930s, farmers increasingly used the disk plow. Its effectiveness mainly depended on improvements in tractors.

The three-point hitch, developed by Harry Ferguson, made it possible to easily attach a variety of tools besides plows to the tractors. The jointer, a miniature plow mounted ahead of the main plow to cut a shallow furrow and clear trash away, had been used in Michigan in the 1840s. Five or more plows might be pulled at a time. The two-way plow had appeared in the 19th century, primarily for use on hillsides; in the early 20th century, it became popular in irrigation farming and was redesigned to be used effectively with tractors.

The subsurface sweep, a plowshare without a moldboard, came into increasing use in the 1930s, particularly in areas subject to wind or water erosion. The increased traction, the high-quality steel, and mass production represented advances, but the sweep design was ancient. The plowing technique of turning a furrow had been developed in medieval Europe in an area not subject to soil erosion. The Dust Bowl of the 1930s and advancing gully erosion in the southern United States brought a revival of the first plow design, now pulled by a tractor.

By the 1920s, several inventors introduced rotary tillers for use in gardens and on small acreages. They consisted of small gasoline engines mounted on two wheels. The tractor moved forward and also forcefully rotated blades, knives, hoes, or tines to cut the ground or clear trash. The first units ranged from three to four feet wide. The tillers were restricted to use in gardens and part-time farming operations on small acreages. Larger machines, which used the same principles and operated from full-scale tractor takeoffs, also came into use after World War I. The rotary tiller pulverized the soil and effectively distributed manures and fertilizers in the soil, but it also laid the soil open to wind and water erosion.

DEVELOPMENT OF COTTON PICKERS

Advances in harvesting methods between 1914 and 1945 consisted mostly of applying tractor power and making larger machines. Cotton, for example, had long been among the crops that could not be readily harvested by machine. Stripper harvesting of cotton began about World War I. Expensive commercial strippers, pulled by mules, first came on the market in 1926. A few single- and two-row, tractor-mounted strippers appeared in 1930. Strippers tore off the entire boll of cotton, ripe or not. Cleaning cotton took place during ginning with uneven results. New Deal support programs and the greater prosperity of World War II apparently caused an increase in the use of strippers in the 1940s.

The spindle cotton picker, after over twenty years of work by inventors, was produced commercially by International Harvester in 1942. It became common in the 1950s and 1960s. Cotton breeders developed varieties of cotton that matured at nearly the same time and grew to uniform size. The spindle cotton picker cut the man-hours for crop production from about 150 per acre to about 25 per acre in most places! Harvesting time fell from 24 to 6 hours on the dry Great Plains.

The spindle picker had a drum which slowly turned as it moved past the cotton plant. On the drum were many small spindles, about a fourth of an inch in diameter and four inches long, which were placed about four or five inches apart. Water was squirted on the spindles so the cotton fibers would adhere to them. The cotton was sucked off the spindles as the drum revolved and was blown into a large cage in back of or on the picker. The cotton was thus picked clean. Most importantly, the spindles would not harm bolls not yet open. After ripening, these could be picked at a later time. Two passes through the field usually did the job. The cotton picker, like the corn picker, needed an additional source of power, not dependent on a ground wheel and forward movement. The tractor, in short, had to come first.

GRAIN HARVESTERS

For wheat and other small grains, the tractor influenced the types and efficiency of harvesters. When World War I began, farmers in wheat country probably favored the header over the binder. The header cut the grain and left it in windrows for later pickup. The binder apparently broke down often. Binders were improved and World War I saw an expansion in their use. In 1919, manufacturers produced only 4,187 headers, but they made 155,466 binders. That same year they also made 58,000 grain cradles. The binder was one of the first machines to be used with the tractor power takeoff in 1919. Farmers

One-row cotton picker, *ca*. 1945. Courtesy International Harvester Co.

in the United States used an estimated 1,250,000 binders by 1938. In 1927, the sale of headers in America had fallen to 1,000, and they had virtually disappeared by 1938.

The binder and the stationary thresher remained the chief machines for harvesting and threshing of the U.S. grain crop from 1914 to 1945. The combine had appeared in the 19th century, but it did not dominate grain harvesting until World War II.

In 1930, manufacturers added pneumatic tires to tractors and the combine found new favor because of the lighter weight and faster operation the tires made possible. Soybeans, which became an important crop in the 1920s, could be harvested with the combine. In 1928, Illinois farmers used 400 of the machines in their soybean fields. In 1937, manufacturers sold more than 28,000 combines. Possibly 15,000 had six-foot wide (or less) cutter bars. The number of combines rose from some 4,000 in 1920 to 90,000 in 1937 during a period of farm depression! The combine made substantial reductions in farm labor possible. The labor cost exceeded the cost of the machine in the long run. The reduction in man-hours ranged from 2 to 5 hours per acre, averaging 3.5 hours. These machines typically harvested an average of 400 acres, but could handle 800 acres or more under favorable circumstances in the 1930s and 1940s. The limit in use generally reflected the size of the wheat farm rather than the limitations of the machine.

Harvesting Kansas wheat, 1945. Courtesy USDA

TRACTOR IMPROVEMENTS

The tractor provided most of the farm power. Several improvements became standard during the Great Depression and World War II. Before 1910, tractors were heavy, weighing around 11,000 pounds and developing one horsepower for every 450 pounds. Lighter weight became the goal of inventors and manufacturers. Enclosed transmissions, replaceable parts, and high-grade steels appeared before and during World War I, although they became commonplace only in the 1920s. By 1925, after extensive experiments, most tractors had four wheels—two driving and two transporting. In 1917, the Fordson—a low-cost, two-plow, four-wheel tractor—was manufactured, weighing only 3,000 pounds. The sale of tractors began to expand, especially in fruit-raising areas. By the 1930s, weight had been reduced to 200 pounds of machine for each horsepower delivered at the drawbar.

For certain types of heavy work the track tractor was used increasingly. The machine cost more than other tractors, and its tracks caused a higher maintenance cost; but it worked well where traction was difficult or where a low profile was needed, as in orchards. Undoubtedly, the track found its greatest use in road building and in heavy construction. It had declined in use on farms before World War II. It reappeared in the form of the bulldozer with a host of uses such as clearing mesquite and leveling fields for irrigation. Its non-farm use, however, surpassed its farm use.

On other tractors, pneumatic rubber tires became commonplace. Rubber tires not only made the machine more comfortable to ride and easier to move, but they also cut fuel needs from to 10 to 20 percent. Repair costs declined because repair of a rubber tire cost less than repair of lugs. Low pressure pneumatic tires for tractors first came on the general market in 1932. By 1935, 14 percent of the tractors manufactured had rubber tires, and this rose to 47 percent by 1937.

In addition to other benefits, rubber tires rolled better, an advantage where traction in dirt was not needed, such as in pulling farm wagons. They could be used on roads, especially for heavy loads on short hauls. The pneumatic tire had considerable influence on tractor engine design by 1940. Gear ratios giving tractors high speeds for highway operation enabled speeds of 12 or even 20 miles an hour. In the early 1940s, rubber-tired wagons for field work became common in spite of tire rationing and a rubber shortage.

The tractor replaced the horse, which meant it had to accommodate horse-drawn implements at first. Most of these had been designed to be drawn at around 1½ or 2½ miles an hour. Redesign of implements allowed faster tractor speeds. By 1940, most tractors did most of their work between 3 and 5 miles an hour.

Obvious savings resulted from larger implements: more plow bottoms, larger shares, greater width of reapers. Greater speed in operation cut the labor time and provided either labor-cost savings or enabled the farmer to increase his cropland with the same labor.

The power takeoff had been developed during World War I, but it became popular only around 1923. It increased the versatility of the tractor. More than any other element, the power takeoff assisted in increasing the speed of tractor operation. By the 1930s, the power takeoff had been used to run spraying and dusting machines and various fertilizer-distributing machines.

Widely used after 1928, the power lift was almost standard equipment by 1945. The tractor operator could lift a plow or other attachments off the ground to make a turn without leaving the machine. The operation was quick, efficient, and not strenuous. Such things as pressure gun lubrication and other easing of maintenance work steadily reduced upkeep and made the tractor increasingly economical as compared to the horse.

INCREASING USE OF TRACTORS

Tractor adoption increased most rapidly during World War I, the late 1920s, and from 1934 to 1937. The tractor made its earliest and most impressive appearance in the wheat and corn regions of the Pacific, the Great Plains, and the prairies. Lightweight, efficient tractors came into wide use by 1925 in the dairy, vegetable, and fruit areas of New York, Pennsylvania, the states surrounding Chicago, and Cali-

Tractors in use, 1922–1945.

FARMALL, 1924

ALLIS-CHALMERS, 1929

CATERPILLAR, 1932

MC CORMICK DEERING, 1935

OLIVER, 1938

CASE, 1942

MINNEAPOLIS-MOLINE, 1945

fornia. By 1930, the greatest use still seemed to be in high-intensity crops usually found near urban concentrations. McKibben and Griffin reported in 1936 on the number of horse and tractor farms by major crops in a sample survey (Table 22.1), which probably reflected the national picture. The introduction of the all-purpose tractor, with its many appendages, had begun to change the pattern in the midst of the Great Depression. The next surge took place during World War II when the workhorse nearly vanished.

Excluding garden tractors, the number of tractors on U.S. farms rose from 1,567,000 in 1940 to 2,354,000 in 1945—a total increase of 50 percent, or 10 percent a year. At the same time, the number of farms declined from 6,350,000 in 1940 to 5,967,000 in 1945—a decline of 6

TABLE 22.1. Number of Farms, 1936

Major Crop	Horses Only	Tractors Only
Corn area	364	483
Cotton area	869	157
Dairy area	360	312
Small grain area	176	640
Vegetable farms	54	42
Potato farms	235	193
Apple farms	108	200

SOURCE: *Changes in Farm Power and Equipment: Tractors, Trucks, and Automobiles,* 1938, p. 24.

percent. This worked out to .24 tractors per farm in 1940 and .39 tractors per farm in 1945.

CORN PICKERS

The invention of a successful mechanical corn picker was slow in coming. Inventors made abortive efforts in the 1870s and 1880s, and some machines were marketed in 1909. Farmers bought a few horse-drawn and some tractor-drawn one-row pickers during World War I, but the mechanism would not work well off a ground wheel.

After improved tractors with power takeoffs were available, two-row corn pickers and huskers were mounted on tractors in 1928. The self-propelled corn picker did not appear until 1946. In the meantime, the weight of the machines was reduced by steady advances in engineering. Stationary husker-sheller machines had been developed at the turn of the century. In the 1930s, someone simply put the machine on a corn picker. However, farmers usually stored corn on the cob, and then they used stationary shellers. The tractor-drawn pickers predominated, mostly because of the difficulty of getting mounted types on and off the tractors.

The number of corn pickers on farms rose steadily through the 1930s and 1940s. Combination corn pickers and shellers rose from 110,000 in 1940 to 168,000 in 1945, contrasted with 10,000 corn pickers in 1920 and 50,000 in 1930. The increase between 1930 and 1940 seems especially impressive since it took place during the Great Depression. The 1940–1945 rise occurred at a time of shortages of farm machinery due to the war. Portable grain elevators were sold on a significant scale in 1934.

FIELD ENSILAGE HARVESTER

The field ensilage harvester, or chopper, combined the cutting machinery of the corn picker with a portable silage cutter. It eliminated nearly all hand labor and reduced the man-hours needed to produce a ton of silage. The first workable ensilage chopper was made in 1915, but it did not become widely used until an improved version appeared in 1928. Pulled by a tractor and working from the power takeoff, it cut the corn, chopped it, and delivered the cut forage to a truck. The ensilage was unloaded onto a conveyor and carried into the silo. A blower system came into use in the 1940s, but belts continued to be most in use through World War II. The ensilage harvester did not make much headway until after the war.

Although expensive, the machine reduced other costs considerably, which may account for its subsequent rapid adoption. The usual method of ensilage involved cutting the corn, taking it to the silo, and running it through a stationary cutter which chopped it up and blew or conveyed the feed into the silo. This method required about 4 man-hours per ton of silage. The ensilage field chopper required only about 1.2 man-hours. By 1960, some 90 percent of the ensilage was harvested by field choppers. This development took place after World War II.

HAY-MAKING EQUIPMENT

In the United States more than half the farmers reported having a mower in 1936, except for those in the western cotton region where only 42 percent owned them. A mower with a five-foot cut was most common. From 1914 to 1936, farmers chiefly used horses to pull the mowers, although some farmers used tractor-drawn mowers by 1936. The major changeover to tractor power occurred during World War II.

During the Great Depression, the side-delivery rake gained in popularity over the dump rake, largely because the side-delivered hay did not have to be teddered. Since the hay was left in a fluffy pile, it also was easier for a hay loader to pick up. However, dump rakes still outnumbered side-delivery rakes two to one in manufacture in 1938.

In 1940, the power-takeoff, side-delivery rake appeared and soon became popular. The side-delivery rake could also be used like a tedder because it could move the windrow over a few feet. Not until after 1945 did the powered side-delivery rake, with a reel to move the hay, dominate.

The hay either had to be stored loose in the hayloft or baled by a stationary baler. Around 1932, the field pickup baler, invented earlier, began to attract attention. This machine saved at least two handlings of the crop. The farmer cut his labor costs again when the self-tying baler appeared in 1940. The machine reduced the number of men needed to handle the baler which had to be pulled by a tractor.

MILKING MACHINES

The milking machine came on the U.S. market in 1905. It saved from 16 to 58 man-hours a year per cow, or an average of 28 hours per cow in a year. One man could take care of milking 6 or 8 cows by 1926. The number of milking machines decreased during the depths of the Great Depression, as shown in Table 22.2. No official census of milking machines on farms was taken until 1940. By 1940, 175,000 farms had milking machines, and the number rose to 365,000 by 1945. The greatest increase took place during World War II.

The Great Depression seems not to have slowed down the innovation and adoption of new devices very much, largely because of governmental aid programs. Viewed against the whole period from 1914 to 1945, the depression years stand out as a period of trial and perfection of machines long dreamed of. The great impetus given by the wars and the economic boom may be deceptive since technological routes out of economic dilemmas were being sought in the late 1920s and 1930s. World events gave farmers the opportunity to profit, but the new technology, like the new economics, favored the larger operations. Through it all ran the story of government aid and the declining number of farms and farmers.

TABLE 22.2. Milking Machines, Percentage in Use on Farms

	1909	1919	1929	1936
		%		
Eastern dairy area	2	11	20	14
Western dairy area	4	18	34	32

SOURCE: R. B. Elwood, A. A. Lewis, R. Struble, *Changes in Technology and Labor Requirements in Livestock Production: Dairying*, 1941, p. 50.

CHAPTER TWENTY-THREE

PRESERVING THE SOIL: GOVERNMENTAL
INTERVENTION ℈ 1914–1945

E ACH TECHNOLOGICAL ADVANCE gave the farmer an opportunity to exploit the land more efficiently and rapidly. Exploitation of land really came down to exhausting the plant nutrients and thus slowly depriving the land of cover. With its cover gone, the land eroded and reached a point where its conservation demanded heroic measures.

As Americans moved westward, they exploited and depleted the soil. Land was cheap and plentiful, but the means for preserving it were scarce and expensive. (Soil erodes fastest where the land slopes steeply, the rain comes in a pelting downpour, the winds blow fiercely, or the land lies flat with few or no trees. Water or wind erosion is the result.) Erosion in the United States took place in ways new to immigrants from western Europe.

In the eastern half of the continent, to the Mississippi and slightly beyond, water erosion posed the greatest danger. Soil exhaustion came first. With humus and trees gone, the soil became like sand and washed away rapidly. Sometimes nearly imperceptible erosion took place during sheet erosion. Successive layers of soil washed away in a gradual process that could take off three or more feet of soil without being noticed—except that the edges of country graveyards had to be walled up to keep them from slipping off onto the surrounding farms. All areas suffered from sheet erosion. More dramatic was the cutting of deep gashes, called gullying. By World War I, the worst damage had occurred in the piedmont of the Carolinas and in areas of Georgia and Alabama.

Wind chiefly caused western erosion. The destruction went fastest during and after World War I when farmers put new land to the plow. By 1930, the most severely damaged places were Oklahoma, Texas, Kansas, and the Dakotas—states in the Great Plains area soon to be called the Dust Bowl. By 1933, at least 50 million acres had been laid waste. Land reduced in usefulness by half through erosion came to another 125 million acres. The destruction had, in fact, just begun.

The damage to the depleted and eroded soil stood out clearly. In wind erosion, the dust and trash smothered grass and crops. In water erosion, the mud and silt clouded and choked streams and silted up dams and reservoirs, increasing the danger of flooding.

THE MUSCLE SHOALS PROJECT

The attack on soil depletion and erosion when it came, came under the guise of national defense. In 1917, a pilot project was undertaken to build a dam for generating electricity and controlling floods. However, the real motivation was to manufacture nitrates for gunpowder. To go back a bit: about 1902, the Germans had discovered a way to produce nitrates by air fixation, a process requiring large amounts of electric power. Nitrates are used in fertilizers and as an essential ingredient in gunpowder. The chief natural source of nitrates was the mines of Chile, on which the United States had become increasingly dependent. When war began in Europe in 1914, the United States realized the vulnerability of the Peruvian and Chilean nitrate supplies that came by way of a long sea voyage. The strategic material could be produced in the United States in large quantities by expanding use of the air fixation method. In 1916, the National Defense Plan called for the building of a federally owned and operated dam and plant for generating electricity to be used, in turn, for making nitrates.

The Muscle Shoals Project consisted of a steam-generating plant and a hydroelectric plant located on the Tennessee River at Muscle Shoals, Alabama. When the war ended the Wilson Dam was not completed, but the steam-generating station was finished. The battle of the private power interests versus governmental interests for control of the Muscle Shoals Project continued until 1933.

During the Harding administration the whole project was considered for liquidation. Henry Ford, among others, offered to buy the facility for a fraction of its cost. Senator George W. Norris of Nebraska led the fight and marshalled the forces to keep the project under federal control. In the Coolidge administration, Congress passed a bill to allow production of power and nitrates, both to be retailed by private industries. Coolidge vetoed it as too radical. A similar plan was vetoed by President Hoover as too socialistic. Meanwhile other problems in the Tennessee valley had become increasingly apparent in the 1920s.

NEED FOR FLOOD CONTROL

The Mississippi had always flooded, but the floods became increasingly severe in the 1920s. They covered more land and recurred with depressing regularity. Countless streams fed into the Mississippi, and the valley of drainage extended from Pittsburgh, Pennsylvania, in the East to Great Falls, Montana, in the West. About half a continent of water drained rapidly to the Mississippi.

In 1927, a new flood crest was reached. It appeared then that the levees, the ancient system of dikes, had reached a critical height. As men piled more dirt or other weight on the levees, the bases washed out. Where they did not entirely collapse they simply eroded and sank. The levees could not be effectively raised at all.

Technologists recognized that the cutting of timber on the headwaters and the plowing of grasslands—such as the prairies of Ohio or the bluegrass regions of Kentucky—had caused the rapid runoff. Reforestation and erosion controls seemed the only long-term solution to the increasingly devastating floods on the Mississippi. The short-term solution called for holding back the floods on the headwaters by a series of dams. Electricity and nitrates could also be produced, especially in the valley of the Tennessee.

The Tennessee River presented problems which could be solved by dams and flood control. Navigation on the long, wandering Tennessee was impossible for its full length. Muscle Shoals, Alabama, was about midriver, where the river spread out over rocks with almost no banks. Engineers and others wanted to dam the river to get it high enough for navigation, and they planned to build a series of locks. In 1930, Herbert Hoover suggested that the government build a series of dams and locks to improve navigation and to raise the river level nine feet. No action was taken.

The Tennessee valley also attracted attention because of the run-down condition of the farms and cities of the area. Erosion was far advanced, and the people had the lowest standard of living in the United States. Poverty in this valley extended over at least seven states.

TENNESSEE VALLEY AUTHORITY

The Muscle Shoals Project became an issue in the election of 1932. Hoover (despite the Hoover Dam built from 1931–1936) opposed the socialism involved. Roosevelt favored the Muscle Shoals Project. In May 1933, Congress set up the Tennessee Valley Authority (TVA) under the New Deal. The TVA was created as a government corporation which would try to solve many problems simultaneously on a regional rather than a state basis. The TVA incorporated the Wilson Dam and Muscle Shoals station.

Private companies wanted to distribute the electricity. Franklin D. Roosevelt wanted government distribution as well as production of the electricity generated. The chief objective of the New Deal was to increase rural electrification, both as a battle against poverty and as an element in the whole problem of fertility restoration and erosion control. Cheap rural electricity would improve rural economic conditions. Counterarguments that private power companies could do the job made little impression on Congress, since the companies had not provided rural electrification. Congress authorized the government (acting through the TVA) to distribute power. However, the TVA had to give retailing preference to rural cooperatives and municipal systems.

Court battles ensued while the TVA corporation went ahead and built a series of dams. In 1936, the matter finally went to the Supreme Court which decided in the Ashwander case that the Federal Government could build dams and distribute power. The Court pointed out previous constitutional interpretations, which gave Congress clear power to improve navigation and control floods. The Court held that power production was only incidental to the main program of the Tennessee Valley Authority.

In 1937, Congress authorized the TVA to buy retailing facilities of private power companies. Others were sold to cities. By 1942, every city in Tennessee owned a system and bought power, and many cities in nearby states did the same. But the main thrust was for rural electrification. The TVA also built rural distributing systems for farmer cooperatives. The cooperatives in turn repaid the TVA for the advanced capital out of the sale of the power. Congress set the electric rates very low, and the TVA had become the largest national distributor of power by 1945.

The center for the unified flood control was the Kentucky Dam. Water was held in the various dams during the high-water season and released downstream to other dams for impounding if necessary. Then, under coordinated control, the water was released in the dry season to get ready for the next year's floods. Navigation automatically improved because the dams kept a good stage of water through the dry season not only on the Tennessee, but also on the Mississippi. The systematic release of water from the Tennessee valley system ended the flood-and-drought cycle on the Mississippi. However, the rate of silting limited the life of each dam.

A coordinated system, the TVA worked well, if sometimes slowly, in aiding farmers in restoring and maintaining soil fertility and halt-

Contour ridging, plowing, and strip cropping, Texas, *ca.* 1940s.

ing erosion. The TVA encouraged reforestation and put much of its own land into forests. The TVA did not begin nitrate production until 1944. The use of fertilizers was encouraged by the development of the superphosphates of the area. The farmer could buy fertilizer for the cost of transportation.

PROBLEMS OF WIND EROSION

The western areas of wind erosion had no such agency, nor could the problem be approached in the same way. The dust storms became progressively worse, with the worst in 1934 and 1935. Something had to be done, and apparently only the Federal Government could do it. Dry farming methods only complicated the problem.

In 1923, some Canadians discovered that planting crops in strips, 40- to 100-feet wide, at right angles to the wind, cut wind erosion on the fallow fields of the dry farming system. The growing grain acted as a windbreak across a wide expanse of land, and when the wind almost touched down to the soil again, it would encounter another strip of growing grain. However, the discovery made little impression on farmers.

For years, conservationists had urged farmers to return some of the plowed land to grass. Some states had tried to encourage this, but reseeding cost too much money. In the 1920s and early 1930s, many farmers lost their farms. The means for recovery were at hand and methods of control known, but lack of capital and ignorance inhibited correction. Effective action awaited the advent of the New Deal in 1933.

CIVILIAN CONSERVATION CORPS

In his campaign of 1932, Franklin D. Roosevelt had proposed the creation of a Civilian Conservation Corps (CCC) to battle fertility depletion and soil erosion. In 1933, Congress established it. Congress put most of the camps under the control of the Department of Agriculture, but some were under the Department of the Interior and the War Department. Congress also set up an Emergency Soil Erosion Service in the Department of the Interior which worked with the CCC.

The corps recruited unemployed single men, 18 to 25 years old, who joined the Civilian Conservation Corps for one year at a time. They received maintenance, food, and pay of one dollar a day, of which they had to allot 25 dollars a month to dependents. The rest of their pay provided enough for recreation and even savings. The men were supported, their families helped, and a host of conservation efforts carried on, most quite successfully. They went to special camps where they worked on improving national forests and national parks. They built roads, firebreaks, and various conservation projects. They worked to check erosion, particularly water erosion, and set up some 41 demonstration projects throughout the country. These showed the farmers what could be done at small expense. The program successfully stimulated public interest in conservation.

SOIL CONSERVATION SERVICE

As one result of the early program, Congress created a permanent agency, the Soil Conservation Service, within the Department of Agriculture in 1935. The same year Congress also extended and expanded the Civilian Conservation Corps. In that year the Department of Agriculture had 2,106 camps; the Department of the Interior, 698 camps; and the War Department, 112 camps; with a total enrollment of 600,000. The War Department used its share mostly in the Corps of Army Engineers. Between them, the Soil Conservation Service and the Civilian Conservation Corps set up over 150 demonstration projects.

The main object was to save the soil. During 1935, the CCC constructed 294,000 small check dams for gully control and seeded 82 million square yards of gully banks. It did an impressive amount of ditching, draining, and damming. Fertility preservation began with crop rotation and manuring. Except for the Tennessee valley, commercial fertilizers were not readily available. Erosion control emphasized contour plowing. Contoured fields first appeared in significant numbers during the 1930s. The Soil Conservation Service advocated strip-cropping for land subject to wind erosion and contour ridging for pastureland.

On the Great Plains—the Dust Bowl of the 1930s—special problems called for special solutions. The Soil Conservation Service developed the programs and the Civilian Conservation Corps demonstrated them. First of all, a land cover had to be planted and maintained. The government distributed some free seeds of native grasses. Sorghum and wheat stubble could be left standing, reducing wind erosion. Lister plowing—deep plowing which produced clods—adequately broke the soil without exposing it to the wind. All of these methods became more common on farms.

SHELTERBELT PROJECT

For more than half a century, one proposed solution to wind erosion had been the planting of trees as shelterbelts. The counterargument held that trees would not grow on the Great Plains. However, the Great Plains lacked trees not so much because of arid conditions as from the ravages of the prairie fires that regularly swept the surrounding grass and destroyed the young trees. Sufficient control of fires with firebreaks, such as several plowed furrows, made it possible for trees to grow.

Although Roosevelt personally urged the development of the shelterbelt project, it ran into continual opposition from Congress. Nevertheless, by shifting the project from agency to agency and administratively ignoring Congress, Secretary of Agriculture Henry A. Wallace managed to keep the program alive. The New Dealers wanted to plant shelterbelts the entire length of the Great Plains and to erect them at intervals from west to east. The belts would run from north to south. The small part of the ambitious program actually accomplished was generally successful.

The shelterbelt idea was probably brought from Russia by Mennonites whose successful experience dated from the days of Catherine the Great. Hundreds of millions of drought-resistant trees were planted under the auspices of the Forest Service and other agencies. The trees were to be planted in rows of six. As the trees grew, they were found to cut erosion. The shelterbelts also markedly reduced the loss of moisture through surface evaporation. Some bureaucratic sabotage helped discredit the program in some places. Farmers eventually cut down many of the trees for fuel, something the planners did not take into account. Still, thirty-five years later many of the trees remained as evidence of a successful beginning. The program continued on a private scale through the 1970s.

Slowly but surely the land was saved. Farming practices changed, never to be reversed. The use of manures and fertilizers became standard for American farmers, and such use steadily increased. Ways were found to make fertilizers, restore soil fertility, halt erosion, cut down on floods, and renew whole areas of the country. And they had just begun. Simultaneously, the Federal Government brought advances to the farmers through large projects such as the TVA and CCC. A long-term relationship between scientists and the Federal Government was beginning. On balance, this relationship has assisted the farmers.

CHAPTER TWENTY-FOUR

GENETICS AND BIOCHEMISTRY:
A NEW AGE DAWNING ⚔ 1914–1945

OOD AND FERTILIZER; breeds and breeding; and biochemistry and genetics dominated the development story from 1914 to 1945. Americans discovered and developed plants and animals that could produce more: an achievement of explorers, of geneticists, and of breeders. Grasshoppers and other insects were brought under control, as were weeds and microbes. The age of DDT, 2,4-D, penicillin, and stilbestrol dawned. Those who first found them thought they promised a happy land of plenty and leisure.

The breeding advances built on the rediscovery in 1900 of the work done by Gregor Mendel, de Vries, and Correns in Europe. Mendel's work showed the way for progress in molecular genetics. His methods inspired others by showing them that they could make genuine scientific discoveries in breeding.

In 1904, an American, George Shull, began studying the fertilization and characteristics of corn. His experiments indicated that self-fertilization and inbreeding of corn yielded a pure strain, but they produced weak plants. He also found that crossbreeding of corn would produce vigorous plants. The discoveries added scientific sanction to a general theory. A better corn was not the goal at first. Corn was used as the chief plant for experiments in genetics. Edward Murray East, whose discoveries paralleled those of Shull, worked at the University of Illinois and then at the Connecticut Experiment Station. He tried to find how inbreeding and crossbreeding worked. In 1910, East went to Harvard. One of his students, Donald F. Jones, got a position at the Connecticut Experiment Station in 1914.

Jones inherited a considerable body of information and a number of hybrid corn types of potential economic usefulness. Obtaining the hybrids seemed to be too complicated for widespread use. Jones decided to see if hybridization of corn could be made economically feasible. He developed a hybridizing technique known as double-crossing, where four pure inbred lines contributed to the final hybrid, instead of the commoner two inbred lines. For many years, double-crossing was the dominant influence on the growing of hybrid corn.

In 1919, East and Jones explained how hybrid corn could be developed. They observed:

> This method is as follows: Four inbred strains are selected which when tested by crossing in all the six different combinations give an increased yield. Two of these strains are crossed to make one first generation hybrid and the other two are crossed to give another. These two different crosses, which are large vigorous plants, are again crossed and the seed obtained used for general field planting. . . .*

East and Jones thought that the system offered better protection for the developer than any imaginable method of patenting. Whoever discovered a hybrid could sell the seeds to farmers and keep the parent seeds and the methods to himself.

Henry A. Wallace, later the secretary of agriculture and vice-president of the United States, was the first commercial producer of hybrid seed corn. In 1926, he advertised the availability of Copper Cross, a hybrid corn, in farm journals. Later that year he formed the Hi-Bred Corn Company, which nine years later was named the Pioneer Hi-Bred Corn Company of Des Moines, Iowa. He pushed his product with vigorous publicity. The high yields on the farm proved the best advertisement. In the fields, as well as in experiments, his double-crossed strains brought an increase from an average of 40 bushels an acre to 100 or even 120 bushels. The economic advantages to the farmer were worth the higher cost of hybrid seed.

Within ten years, hybrid corn had won acceptance by major corn growers. Many companies appeared and improvements continued. The already astounding seed:yield ratio of corn was further improved.

DEVELOPMENT OF IMPROVED WHEAT VARIETIES

Something almost as dramatic happened to wheat, which was mainly grown on the Great Plains by the 20th century. The arid climate made drought-resistant plants essential for successful agriculture. Most wheat resisted drought, but some wheat winter-killed. However, the winter wheats, which lived through the winter, grew again in the spring. The spring wheats had to be planted in the spring for fall. The winter wheats prospered best in the central and southern Great

---

* In Edward M. East and Donald F. Jones, *Inbreeding and Outbreeding* (Philadelphia: Lippincott, 1910), p. 219.

Plains where the winters were not so severe. The spring wheats were better for the northern Great Plains where the winters would kill winter wheats, but the hot dry summers were not so destructive to spring wheats. The different planting and maturation times resulted in a harvesting cycle that began in Texas and ended in North Dakota. From early summer to late fall, threshing and harvesting crews worked from south to north bringing in the grain.

Improved wheat varieties resulted from the immigration of Russian-German Mennonites to the Great Plains from the Crimea and adjacent areas of Russia. Catherine the Great and some of her successors had invited German Mennonites to come to Russia in the hope they would improve the general level of Russian farming. In return, they were guaranteed religious freedom and exemption from military service. But in the 1860s and 1870s, conditions in Russia became increasingly intolerable. Many of these German-speaking Russian Mennonites came to the United States and brought wheat seed with them.

Their wheat, later called Turkey Red, may have been introduced as early as 1860. They indisputably brought Turkey Red to Kansas in 1873. The wheat went unnoticed until the terrible drought of 1887 to 1897. The wheat of the Mennonites still yielded something when other varieties did not. In Kansas and elsewhere, Turkey Red soon became common.

Mark Carleton, a plant scientist for the Department of Agriculture, was assigned to Kansas. Turkey Red greatly impressed Carleton with its high yields and its drought resistance. He convinced his superiors that a search for similiar wheats in Russia and Turkey, for example, might turn up other useful varieties of wheat. He also hoped to find spring wheat suited for more northern regions.

In 1898, Carleton went to Russia and other countries; in 1900, he brought back Crimean, Kharkov, and many other wheats. By 1919, over one-third of all wheat grown in the United States was one of Carleton's importations or a variety developed from his wheat. By 1930, some 40 percent of American wheat acreage was in Kharkov, Crimean, or some variant. One of these variants was Kanred.

The agricultural experiment station in Kansas developed Kanred in 1917. The station released it promptly to farmers. The station also developed Tenmarq in 1917. It was a cross of Turkey Red and Marquis. However, the Department of Agriculture did not release Tenmarq to the public until 1932. In the meantime, scientists used it in crosses to develop other wheats. Among these were the sensationally successful Papooses. These wheat varieties were given the names of Indian tribes who had once inhabited the region. The first of these, Comanche, was developed in 1928, and Pawnee and Wichita came later.

The experiment station did not release the Papooses until World

War II. Comanche came out in 1942, Pawnee in 1943, and Wichita in 1945. All were winter wheats, especially intended for the central and southern Great Plains. In that region, rain, if any, came mainly in the fall and winter. These wheats largely accounted for the astounding increase in wheat production during World War II.

On the northern Great Plains and in the North generally, however, farmers had to grow spring wheats. Two lines of development created wheats suitable for a northern, arid climate. In 1892, C. E. Saunders, a Canadian scientist, developed Marquis wheat which soon was widely used in the Canadian provinces. Marquis did not enter the United States until 1912, and it did not become widespread. While it was drought resistant and high yielding, it shattered during harvesting and, more importantly, it was not rust resistant.

During this time, Mark Carleton had brought back the durum spring wheats from Russia. These were distributed in North Dakota and possibly elsewhere. Durum wheats became popular in the northern Great Plains area. They had two conspicuous advantages: they resisted stem rust very well, and they were rich in gluten. Working in part from Carleton's wheats, geneticists developed Ceres in North Dakota in 1918. Ceres added to the overproduction of the 1920s and 1930s.

C. B. Waldron of the North Dakota experiment station had successfully crossed the Canadian Marquis with the Russian Kota to make Ceres. The station released it to farmers in 1925, and they grew it widely in the spring wheat area by 1930. In 1934, Thatcher wheat was also developed out of Marquis. These were hard wheats—suited to the new milling process because they did not crumble. All were prized by bakers and other consumers. The disaster of the Great Depression and of the drought of the 1930s would surely have been worse without these new wheats. The fantastic production achievements during World War II certainly owed much to the new wheat varieties as well as to improved methods.

INCREASED USE OF SORGHUM

The depression and drought brought a renewed interest in another grass, sorghum. An xerophyte—it not only survives drought, but actually flourishes under conditions of water scarcity. Sorghum's waxy leaf surfaces resist transpiration; under conditions of extreme water shortage, the plant merely stops growing until moisture becomes available again. Other plants generally die when deprived of water for any length of time. The ground seeds make good feed, and the plant makes excellent fodder. Cattle like it, possibly because of its large sugar content. However, the green plant contains large amounts of hydrocyanic acid which poisons livestock if fed green. The poison may explain its relative immunity to insect attack. Farmers and stockmen found that in years of drought and insect attacks, often only the sorghum crop survived.

Sorghum originated in Africa, and apparently slaves brought sorghum seeds to the United States. The varieties ranged from sorgo, or sugar sorghum, to milo, kaffir grass, broom grass, and over a thousand other types. By the 1870s, it was well established on the southern Great Plains, and it became increasingly popular among livestock raisers. However, the plant did not adapt well to northern climates. Around 1877, scientists in Minnesota developed Minnesota Amber which survived northern climates and managed to reach maturity as well. Many more varieties were adapted to the northern climate, but full adaptation on the northern plains did not occur until 1915. Sorghum had attracted so little interest by 1930 that when farmers needed it, seed for many varieties could not be found in quantity.

The Dust Bowl farmers discovered that sorghum stubble would hold the soil, and sorghum grew when nothing else would. In South Dakota, for example, the amount of sorghum harvested rose from 15,655 acres in 1929, to 472,890 acres in 1934, and 1,071,895 acres by 1939. After 1934, sorghum's success helped reestablish animal husbandry on land which suited animals better than plants. The development of ranch farming owed a great deal to sorghum.

Similar developments of new strains and the introduction of new plants took place in other areas and crops between the wars. Long-staple, dryland cotton was developed for the southern Great Plains, range grasses came from Russia, and soybeans entered from Asia. Experimenting and improvement went on steadily.

IMPROVEMENTS IN LIVESTOCK

Similar advances took place in livestock genetics. In 1934, Danish Landrace hogs were introduced into the United States in an effort to develop a meatier hog. Scientists subsequently developed others, such as Minnesota #1. The Danish importation marked a new interest in hog breeding. In 1936, the Swine Breeding Laboratory was established at Ames, Iowa. Breeding experiments were speeded up at the Beltsville Station (Maryland) of the Department of Agriculture. Improved animals that could produce more were increasing before World War II.

About this time, artificial insemination for dairy cattle came into use for the first time in the United States. The advantage of artificial breeding was that bulls of proven ability to sire good dairy cows could be used by many dairy farmers. A bull could service 30 to 50 cows a year; but with artificial insemination, it could service at least 2,000. In 1938, the farmers began the first artificial breeding association in New Jersey. The system soon spread to other dairy regions. The number of artificially inseminated cows rose from 7,539 in 1939 to 1,184,000 in 1947, or 4.6 percent of all dairy cows in 1947. In 1937, the average U.S. cow produced 4,366 pounds of milk; this rose to 5,007 pounds per year in 1947.

As feed production rose, due to new plants and methods, animal productivity increased. Food for animals also became more abundant simply by the increase in the use of gasoline-powered machinery and the decrease in horses and mules. Between 1920 and 1946, the savings in feed for horses and mules would have fed 32 million hogs to market weight.

CONTROL OF GRASSHOPPERS

Losses to insects involve every conceivable plant and animal. For drama and for comprehensiveness of victims, however, nothing quite equaled devastation by locusts. Such plagues had occurred regionally at intervals throughout history in the United States and elsewhere.

Efforts to control the insects went on year after year. Indeed, earlier plagues had brought about the formation of the Bureau of Entomology in the Department of Agriculture in 1877. By the turn of the century, scientists and technologists were convinced that the best method of attack against grasshoppers consisted in spreading poison bran in areas of concentration before the grasshoppers (also known as locusts) became migratory. Once they became migratory, no one could do much about them.

The poison bran method of attack involved the identification of the area of probable outbreak and the spreading of poison. This called for more organization and expertise than was commonly available at the site of an outbreak. Defensive measures called also for more money than a community could usually muster. State and federal cooperation for grasshopper control began to develop in the 1920s. The program worked well and served as a precedent for the terrible plagues of the 1930s.

The first outbreak of serious proportions developed in South Dakota, Nebraska, and Iowa in 1931. More localized hordes came in 1932 and 1933, increasing in 1934. Destruction could be 100 percent on a vast acreage. Damage increased in 1934, decreased in 1935, and then hit an unbelievable peak in 1936. Losses that year to locusts

Grasshopper-eaten cornstalk, North Dakota, July 1936. Courtesy Library of Congress

came to at least \$106,333,000—more than the gross farm income for Arizona, Nevada, New Mexico, Utah, and Wyoming combined. By destroying even stubble, the locusts laid the land open to wind erosion, contributing substantially to the Dust Bowl. The drought obscured the work of the insects which, nevertheless, did terrible damage on their own.

The last great locust plague hit in 1939, when grasshoppers stripped Colorado, Wyoming, Nebraska, Iowa, South Dakota, Montana, and North Dakota clean. Grass, corn, wheat, vegetables, tree leaves, and even clothing hanging on lines disappeared before the voracious hordes. In many places within the area of attack, devastation was virtually entire. Cars overheated because of clogged radiators, and railroad engines stalled because dead insects kept the wheels from getting traction on the rails. Nothing like that should ever recur because of the discovery of inorganic insecticides, of which DDT led the way.

DEVELOPMENT OF INSECTICIDES

A Swiss company discovered the insecticide DDT and patented it in 1939. By 1943, it was available in the United States. Its first spectacular success was in 1944 when the army used it to combat lice and typhus in American-occupied Italy. This chlorinated hydrocarbon controlled a wider range of external parasites than any other known insecticide. It could be used as a dust, a spray, or a dip for plants or animals. Although it was not used on grasshoppers in the United States to any extent, DDT was the precursor of a long line of hydrocarbon insecticides that did effectively bring the grasshopper to heel.

Other insects also succumbed to the new insecticides. The savings are inestimable, but unquestionably they are vast, and insecticides also contributed to the astounding increase in agricultural productivity after World War II. The giant steps in this phase of biochemical activity took place during the war, and the effects came after the war.

DISCOVERY OF HERBICIDES

Another miracle of biochemistry was an outgrowth of research in chemical warfare during World War II. Most of the work took place at Fort Dietrich in Maryland in the 1940s. True herbicides were first developed there.

The work had not begun as an effort to find a weed killer. Actually scientists stumbled onto it as they studied plant growth hormones. In 1935, a hormonelike substance was synthesized, and at least 54 substances that affected plant growth had been found by 1939. That same year, other scientists discovered synthetic hormones could make ripened fruit hang onto trees longer and hasten ripening and coloring of fruits.

Crop dusting, June 1942. Courtesy Library of Congress

E. J. Kraus of the University of Chicago thought the hormones should have other uses. He noted that too much of the various hormones injured and even killed plants. In 1941, Kraus suggested the synthetic hormones might make good weed killers. In June 1941, R. Pokorny reported the synthesis of 2,4-dichlorophenoxyacetic acid (shortened to 2,4-D in 1945); and in April 1942, Zimmerman and Hitchcock reported its strength as a growth regulator.

Kraus directed experiments on herbicidal use of 2,4-D. He discovered the hormone was toxic to broad leaf plants and accidentally found it also would kill rice. By June 1944, John W. Mitchell and Charles L. Hamner recommended 2,4-D as an effective herbicide. The chemical was marketed in 1945, and a new age of weed control dawned for the U.S. farmer. The backbreaking work of hoeing and the expense of cultivating were destined to end. As in the case of DDT, 2,4-D had its effect after the war that had hastened its discovery.

CONTROL OF BOVINE TUBERCULOSIS AND BRUCELLOSIS

Diseases of animals caused annual losses of depressing proportions. As work went on to improve livestock, it also went forward to bring diseases under control. The attack took two forms: cure when possible, otherwise elimination of the diseased animals. Killing diseased animals was easier than curing them. However, the discovery that microbes could cause disease opened the way to control of disease with immunization and medication.

Searching for a way to immunize against tuberculosis, Robert Koch of Germany accidentally developed the tuberculin diagnostic test in 1890. This test could show when people or cattle had tuberculosis, even if they seemed healthy. By 1892, the first tests were made on American dairy cows. By 1900, the tests had shown such an alarming incidence of bovine tuberculosis in U.S. dairy herds that many farmers refused to have their cows tested. Many asserted humans could not contract bovine tuberculosis, a claim later shown to be inaccurate. Some public health officials and social workers estimated that in the larger cities bovine tuberculosis accounted for some 65 to 85 percent of all known tuberculosis cases in 1900. They observed also that bovine tuberculosis generally attacked glands, bones, and other organs; human or aviary tuberculosis was usually a respiratory illness. The clinical observations and distinctions were imperfect, but they turned out to be accurate enough.

Publicity about bovine tuberculosis resulted in rising demand for the eradication of the disease in dairy cattle. The only feasible method seemed to be the slaughter of all diseased cattle. Some states undertook testing programs, but they met with farmer resistance and violence, even when the states partly compensated the farmers for the destroyed cattle. The whole matter seemed stalemated in 1917, when the Federal Government began a program of testing and compensated slaughter. Actually, the Federal Government paid only a portion of the value of the slaughtered cow; the state had to pay a part, and the farmer had to bear a small loss.

In 1917, the Federal Government also initiated the area plan for the testing and eradication of bovine tuberculosis in the dairy states and ultimately the entire country. The results were startling in terms of improved human health—to say nothing of cattle health and productivity. The decreasing death rates due to the different types reflected not only better care for victims of the disease, but they suggested positive results for the program of eradication of bovine tuberculosis.

Thanks to the federal program, the United States was virtually free of bovine tuberculosis by 1942. Although the disease had not been completely eliminated, it had been brought under control. Systematic rechecking would keep it that way, and the disease almost disappeared.

Fear of tuberculosis increased public interest in and demand for pasteurization of city milk supplies. The demand centered in the larger urban areas and took the form of city ordinances. Effective machines for pasteurizing milk came on the market in 1895, and that same year a steam turbine bottle washer was invented. Some 874 cities had compulsory pasteurization laws by 1927. These cities thus effectively required the pasteurization of most U.S. milk.

Pasteurization aided the consumer, but it did little to improve the health conditions for farm animals. In fact, some critics called pasteurization the "dirty milkman's makeshift." As for economic importance, urbanites showed their confidence in getting clean milk by consuming greater amounts of it per capita. Since healthy cows give more milk longer, the farmers also benefitted directly from the tuberculosis control program and the enforced pasteurization.

The other long-term problem was brucellosis—also known as contagious abortion, Bang's Disease, Malta Fever, and undulant fever. Not until Alice Evans of the Department of Agriculture proved it in 1915, did anyone know for sure that undulant fever and brucellosis were the same disease. The disease caused abortion in livestock and in humans. Sometimes it caused death; more frequently it reduced productivity.

Until World War I, brucellosis seemed to be a problem for veterinary science. Then it also became a problem for human medicine. For consumers, the danger could be averted by pasteurizing milk. Until after 1945, only the segregation of diseased animals and scrupulous cleanliness worked very effectively. Even so, many a dairy farmer and his family contracted the painful disease.

Like men, animals contract a multitude of diseases caused by various agents. In many infectious diseases, animal care follows the same pattern as human care. Since scientists perform their experiments on animals first, animals sometimes turn out to be the first beneficiaries. On the other hand, since people eat meat and drink milk, medicines which leave a residue in animals sometimes cannot be used.

DEVELOPMENT OF PENICILLIN AND SULFA DRUGS

Alexander Fleming of England discovered penicillin in 1928. But the sulfa drugs first made an impact on clinical medicine in 1935. Sulfanilamide was the first to be synthesized, and it proved quite effective in the control of bacterial infections. The sulfa drugs were very useful in saving injured animals, but they had little effect on prolonged sickness. The bacteria became resistant to the drugs in extended treatments. A long illness used up its energy resources so that the animal could not muster enough additional resistance to effectively use the drugs. The kidneys of animals were affected, especially in prolonged treatment, and the sulfas were also excreted in milk. The latter effect had considerable importance for human consumers.

Sulfa could be put into feed and water easily. The ease of supplying the medication was important in treating groups of animals. A herd or flock could all be treated at once. Well before World War II, the sulfa drugs had reduced disease losses for many farmers, especially poultrymen.

Then came another breakthrough with penicillin. H. W. Florey and E. B. Chain restudied Fleming's discovery ten years after he made it. All three of them received the Nobel Prize in medicine in 1945. (Koch had received it for his tuberculin test in 1905.) Penicillin was the natural product of a mold that resisted efforts to speed its production. As late as 1943, very little penicillin was made, but American engineering ingenuity found ways of increasing the supply. Soon single animals received doses as great as the entire supply available to Fleming in 1928. Penicillin and subsequent antibiotics greatly reduced livestock death losses. However, antibiotics lost so much potency when fed that for many years farmers seldom used this method of medication.

Research had produced higher-yielding breeds of plants and animals and ways of cutting disease losses. The search for new sources of food produced little effect at first.

DEVELOPMENT OF OLEOMARGARINE

But one fairly successful effort at inventing a food took place—the manufacture of oleomargarine (also called oleo or margarine). The original process was invented in France by Hippolyte Mège-Mouriés, a chemist who patented the method in 1869. The butter substitute used lard and tallow which might not otherwise have been used as food. The chemical process gave these fats the consistency of butter and removed the rank taste of the original fats. Often the product was mixed with butter, which was better than the earlier and cruder mixture of lard and butter.

Apparently the earliest manufacture of margarine in the United States was in New York in 1872. The product attracted enough hostility from dairymen to be punished with its first federal excise tax in 1888 and the more important Grout Tax of 1902. Congress amended the Grout Act several times, usually upward, placing the highest taxes on colored margarine. Dairymen had long been interested in the science of nutrition. The dairy interests subsidized research at the University of Wisconsin, which they hoped would show the nutritional inferiority of oleomargarine.

DISCOVERY OF VITAMINS AND THEIR USE IN OLEOMARGARINE

From the turn of the century, scientists had found increasing evidence of some mysterious substances in food necessary for animal health. In Europe, Casimir Funk, working on the problem since at least 1911, discovered the presence of something he called a "vitamine." Elmer V. McCollum at the University of Wisconsin took up this general line of research, and in 1915 he announced the discovery of fat and water soluble vitamins which he later called "A" and "B." In the process he showed by experiment that oleomargarine was badly deficient in these substances. Vitamins would have been discovered anyway, but the battle between margarine and butter had been a spur to the discovery which probably came sooner because of the contest.

Oleomargarine soon was vitamin-enriched and thus was as nutritious as butter. In the United States, butter steadily lost ground to oleomargarine in sales (Table 24.1).

The first substantial gain for oleomargarine came immediately after World War I, possibly due to a depression then. During the Great Depression, margarine made steady but unspectacular gains.

TABLE 24.1. Consumption in Pounds per Capita

Year	Butter	Margarine	Year	Butter	Margarine
1910	18.1	1.2	1930	17.2	2.6
1915	17.0	1.4	1935	17.1	2.9
1920	14.6	3.4	1940	16.9	2.4
1925	18.0	2.0	1945	10.9	4.0

SOURCE: *Agricultural Statistics* (for relevant years after 1935).

The 1940 figure reflects wartime prosperity and wartime shortages. During those years both butter and margarine consumption went down. In the 1920s and 1930s, manufacturers (especially Lever Brothers) found ways of making margarine from vegetable oils, and they added vitamins. Any onus that margarine had faced as a form of chemically reconstituted lard thus tended to disappear.

The big break for margarine came when the war brought food rationing. Consumers had red stamps as rationing currency with which they bought meat, butter, and margarine. What the consumer saved on butter rationing points by buying margarine he could use for the more expensive meat. Margarine held and steadily increased the market achieved during the war.

The movement from animal fats to vegetable fats in margarine during the 1930s markedly changed farmer relationships in the battle against repressive taxation. Cottonseed oil, soybean oil, and imported vegetable oils all suffered from the oleomargarine taxes. Slowly states eliminated repressive taxes: first in the cotton South and then in the soybean and corn states of the North. Pressure mounted for the end of the punitive federal tax, which came only after World War II when most people ate far more margarine than butter. By 1950, a sizeable portion of the population had become fed up with the dairyman's law.

Oleomargarine made vegetable oils, including coconut oil, palatably available. The increased supply of oleomargarine hurt the butter industry in America. Dairy farmers, however, sold more ice cream, cheese, and fluid milk. The use of oleomargarine thus brought an absolute increase in the food supply and a lessening of costs.

The ultimate impact of the discovery of vitamins was tremendous. Shortly after the discovery of vitamins A and B, Charles King found vitamin C in 1919. Scientists not only isolated them, but synthesized them eventually. Processors could add vitamins to foods. Not until about the 1940s did commercial feed companies systematically add vitamins to animal feeds. In combination with other additives, they markedly increased livestock production. The greatest impact came after 1945.

Biochemistry and genetics had produced many changes even by

1945. Genetics led to hybrid corn, new varieties of wheat, and better sorghums. Livestockmen benefitted from breed improvement in hogs and cattle. Insects, particularly grasshoppers, came under better control as hydrocarbon insecticides appeared. The ancient curse put on "the-man-with-the-hoe" lifted with the discovery of 2,4-D. Koch with the tuberculin test and Fleming with penicillin both won Nobel Prizes, forty years apart but well within the span of a generation. These discoveries in medicine brought help to animal raisers as well as to mankind in general. Scientists and farmers had produced some wonders with many more to come.

CHAPTER TWENTY-FIVE

THE POSTWAR CONFLICTS ✠ 1945–1972

T HE END of World War II was the beginning of a new era. Farmers, above all else, wanted to avoid the economic disaster that had befallen them after World War I. Nearly everyone expected a depression to follow World War II, although politicians and businessmen took preventive steps. In fact, the disposable income of Americans continued to rise steadily in spite of inflation and higher taxes. Total disposable income per capita in 1964 dollars, with the inflation squeezed out and taxes deducted, showed a fairly consistent rise. The depression and prewar averages show an interesting contrast (Table 25.1).

As the population grew, especially in the lower age levels, per capita income figures concealed some real per family advances. The proportion spent on food declined steadily. In 1950, an average American spent 30.6 percent of his income for food. This fell to 26.6 percent in 1960 and to 17 percent by 1969, in unbelievable contrast to the rest of the world.

American farmers produced more than Americans could possibly consume. As had happened at the end of World War I, surplus U.S. farm products were moved onto the world market in trade or in some

TABLE 25.1.   Income

Year	Average, per Capita	Total Disposable Income
		bil $
1929	$152.8	1.254
1933	116.0	.923
1940	173.0	1.309
1945	240.3	1.717
1950	260.9	1.720
1955	309.0	1.870
1960	360.0	1.993
1964	402.6	2.136

SOURCE: *Statistical Abstract*, 1965.

other way. War loans and relief took up the farm surplus from 1945 to 1947, but the predicted depression seemed imminent by 1947.

Then General George C. Marshall, secretary of state, proposed a gigantic relief program for Europe and the Free World. Congress promptly enacted the Marshall Plan which chiefly involved federal purchase of commodities for the use of other countries. The Marshall Plan amounted to an export subsidy on a grand scale. Food ranked high as an export commodity. The program helped to restore the ailing European economies; it gave farmers subsidized prices; and it fed millions.

In the same year the Marshall Plan began, Branch Rickey of the Dodgers hired Jackie Robinson as the first black major league baseball player. The walls of racial exclusion began to crumble as the United States continued its cold war against Communism. In 1948, the Berlin Airlift became necessary. President Tito of Yugoslavia successfully defied the Soviet Union and sent his nation on its own route of Communism. The Soviets backed down from a military confrontation and the Berlin Airlift ended in 1949. In that year the Marshall Plan also ended. Many of those who had supported it had done so only to contain Communism in Europe. The European Communist threat now seemed less ominous and the Marshall Plan died.

A severe general recession followed immediately. No part of the economy suffered more than the agricultural sector, but the distress proved only temporary. In 1950, the Korean Conflict began. This undeclared war was no trivial enterprise. The increased armed forces brought more food consumers under the protective federal wing. The demand for agricultural commodities increased and prices rose again. The war also produced an increase in per capita income with a better distribution of income as well.

Dwight D. Eisenhower ran for president with a major commitment to end the war in Korea as promptly as possible. He won, took office in 1953, and that year the war ended. The economy once again began to falter, and a farm recession was under way by 1954. The next year farmers in the Midwest (chiefly in Iowa) formed the militant National Farmer's Organization (NFO). They declared wars on processors and announced a coming struggle of major dimensions. Under NFO strategy, farmers would refuse to sell their food unless the middlemen met the farmers' price level, but the final showdown was delayed.

CHANGES IN FARM POPULATION AND PRODUCTION

Farm population declined as consumers continually increased in numbers and urbanization continued. The brief period of peace from

TABLE 25.2. Decline of Farm Workers*

Year	Total Farm Employment	Total U.S. Population
	mil	
1945	10.0	139.9
1946	10.3	141.4
1947	10.4	144.1
1948	10.4	146.6
1949	10.0	149.2
1950*	9.9	151.7
1955	8.4	165.3
1960	7.1	179.9
1965	5.6	193.7

SOURCE: *Agricultural Statistics,* relevant years.
    * Five-year intervals hereafter do not distort the reality.

1946 to 1950 slowed the movement from farms, but thereafter farm workers relentlessly decreased (Table 25.2).

As demand for agricultural commodities slackened after World War II, total production in many of them also declined in response to federal programs and to the market. Wheat production went from 1,032 billion bushels in 1944, to 1,006 billion bushels in 1949, and to 908 billion bushels in 1954. But increased hog production and the larger size of the hogs marketed indicated higher meat consumption and presumably a higher standard of living for Americans. In 1944, farmers produced 83,741,000 hogs which yielded 20,755,000 pounds of pork. The number of hogs fell dramatically in 1949, but the amount of pork produced did not fall as much. In 1949, U.S. farmers produced 56,257,000 hogs to yield 19,457,476,000 pounds of pork. By 1954, the amount of pork fell to 18,218,278,000 pounds from 57 million hogs.

Random indicators suggested a highly complex and even contradictory national economy. The number of tractors on farms increased from 2,354,000 in 1945 to 4,625,000 in 1965 in spite of the decreasing number of farms and farmers. What happened to Americans in general also influenced the life of American farmers. In 1940, the number of unemployed was 8,120,000, or 14.6 percent of the work force; and the number of telephones came to 21,928,000. In 1950, the unemployed came to only 3,142,000, or 5.0 percent of the work force; and telephones numbered 43,004,000. In 1960, the unemployed amounted to 3,931,000, some 5.6 percent of the work force; and telephones totalled 74,057,000. Although not all telephones were in private homes, the figures reflect more than aggressive salesmanship on the part of telephone companies. Statistics can only hint at the good American life, but farmers shared in it—telephone and electric lines crisscrossed all of America during these years.

TABLE 25.3. Growth of Black Population in Selected U.S. Cities, 1940–60

Year	City	Total Population	Black Population	Per-centage
1940	New York	7,454,995	458,444	6.14
	Baltimore	859,100	165,843	19.30
	Chicago	3,396,808	277,731	8.17
	Los Angeles	1,504,277	63,774	4.23
1950	New York	7,887,380	749,080	9.49
	Baltimore	1,330,875	263,785	19.82
	Chicago	5,476,490	586,655	10.71
	Los Angeles	4,357,320	218,130	5.10
1960	New York	10,695,963	1,224,590	11.40
	Baltimore	939,024	325,592	35.80
	Chicago	6,220,913	889,961	14.20
	Los Angeles	6,746,356	464,112	6.80

SOURCE: U.S. *Census Reports,* relevant years.

BLACK MIGRATION TO CITIES

Simultaneously, something amazing happened to the southern blacks from 1945 to 1972. They moved North in unprecedented numbers and to southern cities, changing forever the southern style of agriculture. As in previous migrations, the blacks bettered their lot and provided a comparatively rich urban market for the farmers who remained. The issue of race obscured the essential fact of a black migration to the cities (Table 25.3).

Most of this population increase came from migration, not births. The migrants had better nutrition and better medical care. Reformers found poverty and starvation chiefly in the rural areas. The results showed up both in total population and in the black population at large (Table 25.4).

Women still outlived men of their own race, but nonwhite women had lower life expectancies than white men. In percentage of improvement, the nonwhite population was coming closer to parity with the whites. Certainly a better diet played an important role,

TABLE 25.4. Expectation of Life at Birth, by Color and Sex

Year	White Male	White Female	Nonwhite Male	Nonwhite Female	Color Gap Male	Color Gap Female
1939	63.3	66.6	53.2	56.0	10.1	10.6
1944	64.5	68.4	55.8	57.7	8.7	10.7
1949	66.2	71.9	58.9	62.7	7.3	9.2
1954	67.4	73.6	61.0	65.8	6.4	7.8

SOURCE: *Historical Statistics,* 1957.

TABLE 25.5. Income Sources

Year	Total National Income	Old Age and Survivors	State Unem- ployment	Veterans Benefits
			bil $	
1950	228.5	1.0	1.4	4.9
1955	310.2	4.9	1.4	4.3
1960	401.3	11.1	2.8	4.5
1963	463.0	15.3	2.8	5.0

SOURCE: *Statistical Abstract,* relevant years.

but by 1954 nonwhites had achieved only the life expectancies already attained by whites in 1934!

The income that helped produce the change came from a variety of sources other than wages (Table 25.5). The figures for veterans are most conspicuous. The unemployment benefits figures reveal something about the society and its distribution of economic goods when compared with the unemployment figures (Table 25.6). When the number of unemployed dropped between 1950 and 1955, a drop both absolute and in percentage, the amount of unemployment funds spent did not drop. While the 1960 unemployment figures were only slightly above those of 1950, the payments for 1960 were double those for 1950. The pattern of redistributing urban income helped the farm market. The migration greatly increased the number of consumers of farm products, apart from any natural increase in population.

IMPACT OF WORLD CONDITIONS ON U.S. AGRICULTURE

Another influence on the farm economy arose from world conditions, which fostered profitable disposal of farm commodities. Americans believed an empty stomach conditioned men to accept or even welcome Communism. First they sent relief, and then the Marshall Plan siphoned off farm surpluses. Americans believed their food re-

TABLE 25.6. Unemployed in Total Work Force

Year	Number	Percent
	thou	
1950	3,351	5.3
1955	2,904	4.4
1960	3,931	5.6
1963	4,166	5.6

SOURCE: *Statistical Abstract,* relevant years.

TABLE 25.7. Persons Supplied with Food by One Farm-worker, 1945–65

Year	At Home	Abroad	Total
1945	12.87	1.68	14.55
1950	13.79	1.68	15.47
1955	17.32	2.17	19.49
1960	22.30	3.55	25.85
1965	30.79	6.23	37.02

SOURCE: USDA, ERS, *Statistical Bulletin 233*, 1966.

sources represented a strength that could be used in the cold war against Communism.

From 1945 to 1972, the United States was almost constantly involved in warfare: the Korean Conflict from June 1950 to July 1953; an invasion of Lebanon from July to October 1958; and an invasion of Cuba (the Bay of Pigs fiasco) in April 1961. Meanwhile small numbers of troops had been sent to Vietnam where another undeclared war ran from April 25, 1961, to January 23, 1973. Americans fought these upopular wars with only a small part of the nation's wealth and manpower. Prosperity, growth, and inflation resulted—and an ever-ready market for farm produce. The achievements of the American farmer as a food producer were important. He fed more Americans as well as an impressive number of foreigners (Table 25.7). The number of persons fed abroad reflected not only the success of U.S. farmers, but also public policy. A high rate of inflation and a high standard of living raised the price of U.S. farm products so high that they were not competitive on the world market.

FEEDING AMERICA'S POOR

Around 1965, Americans had their attention drawn to the fact that not all Americans had enough to eat. The food stamp plan was begun in 1965 and later expanded. Surplus foods were shunted into the rural and urban areas of poverty, but this happened more effectively in the urban areas than in the rural areas for a variety of reasons.

Problems of communication were most obvious. To dramatize the plight of the rural poor, the Reverend Ralph Abernathy organized a campaign in Washington, D.C., in the summer of 1968. The poor came mostly from the rural South. Nothing much could be done on such short notice, but the hungry had been found. The cost of going to their local county seats made it impossible for the very poorest to get available help. Not until late in 1970 did it occur to someone to mail food stamps to rural recipients.

Unquestionably, cities offered better distribution facilities, better exposure of needs, and more kinds of assistance. The cities sometimes offered better chances for work. For all the complaints about relief going to the unworthy poor, the farmer actually seemed to be the chief beneficiary. The poor spent a large proportion of their relief money on food. The proposal for revenue sharing, whereby the Federal Government would give the states part of the federal revenue, threatened to put the farmers in a position of supporting their own aid program by their taxes. The war against poverty came last in time and interest. By 1970, some people had already declared poverty the winner, but the war was fought with deadly seriousness by those few in the field of battle.

President Harry Truman had feared a postwar depression, and, to some extent, so did his successors; yet the depression never came. Relief programs in Europe and the Marshall Plan put off the depression, and then came the Korean Conflict. The discovery of a marginal few who could not make it was proclaimed by John Kenneth Galbraith in *The Affluent Society*. In the meantime, the blacks surfaced and their movement from the rural South to the cities accelerated after 1945. In the cities, blacks demanded and increasingly got equal treatment with whites.

Farmers provided needed commodities, and they also became important beneficiaries of social progress. On the whole, the farm community improved its lot in relation to the rest of the country. The farmers were obviously an important part of the American economy. They had always been, but they also had surfaced as the blacks had done. Although ever smaller in number, farmers approached that stage where they might organize to protect their economic interests. At the same time, farming had become so incredibly complex that the professional farmer could no longer be displaced by the amateur raiser of food. The farmers were important, they were powerful, and they would not be quiet.

CHAPTER TWENTY-SIX

FOOD FOR PEACE: THE
NEW MARKETS ☙ 1945–1972

ONSIDERED as a single industry owned and operated by farmers, agriculture ranked first in the United States from 1945 to 1972. It employed more people, and it had a greater capital investment and cash flow than any other industry. In terms of money flow, the second largest industry was philanthropy in its various aspects. Charity represented huge expenditures of money, goods, and time.

Admittedly, some of the philanthropy was suspect. Congress finally recognized this, and the Tax Reform Act of 1969 controlled some of the largest foundations. Still, the Ford Foundation and smaller foundations such as Rockefeller, Guggenheim, Duke, and Carnegie pumped billions into the economy. Churches and universities ranked high as recipients of this generosity. Many owned large amounts of tax-free property which was also classified as philanthropic.

Americans, always a generous and openhanded people, became accustomed to giving to charity in the years after 1945. Giving became almost a national obsession, and it was big business. It ranged from a Ford Foundation grant of several millions to a housewife's donation of a dollar to the local Cancer Society.

Thus Americans did not find governmental charity on a global scale so very strange nor governmental giving as much more complicated than a kind of economic activity with forms and rewards of its own. Because of the farm price support system of the Federal Government and the nearly complete cartelization of farming, even the market had an aura of philanthropy. Many Americans felt they could fittingly use food to do good. Such charity meant in practice rewarding the good and punishing the bad. The good, by definition, opposed the Communists.

In 1945, the United States cooperated with the United Nations Relief and Rehabilitation Administration (UNRRA), which gave away food chiefly in Europe to enemies and allies alike. European agriculture recovered more slowly than it did after World War I. The destruction had been more severe, and modern commercial farming required machinery which had either not been made or had been destroyed and took time to replace. Probably the European drought of 1946 and 1947 delayed recovery more than anything else. The need for American generosity continued and also kept the market alive for the American farmer.

The food program provided a precedent for the large-scale charity in the form of the Marshall Plan. About 38 percent of the assistance under the act went directly for food and the rest for reconstruction. But when the need for massive infusions of food into Europe ended so did the Marshall Plan in 1949. However, the Korean Conflict took up the slack until 1953. By 1954, agricultural surpluses appeared on a large scale that seemed absurd when most of mankind still suffered from hunger. The solution seemed almost too easy.

PUBLIC LAW 480

In 1954, Congress passed the Agricultural Trade Development and Assistance Act, known thereafter usually as PL 480. Public Law 480 in three (later four) distinct programs undertook to feed the world's hungry and to solve the problem of U.S. farm surpluses at the same time. The law prohibited sending food abroad if it was needed at home. Only surplus commodities could be moved under the various titles of the act.

Title I allowed for formal arrangements between the United States and other nations, whereby the United States sold food to the other nations in their currencies. Each transaction called for a formal agreement, but not a formal treaty. The foreign government disposed of the food as it wanted, usually by resale to the distributors already in the country. The Federal Government bought the food in the first place from the farmers directly or from the federal Commodity Credit Corporation which had surplus stocks on hand.

The receiving government might not allow much of its currency to leave the country, and often the United States had no way to use it. So the Fulbright Fellowships and other similar programs came into being. This rather complicated form of charity brought a good many scholars to and from Europe.

Title II provided for assistance with outright gifts in time of disaster or other emergency. Such food could also be used in foreign school lunch programs, in lieu of wages, or for parts of wages for various projects. Title III also allowed some food to be given away except that the distributing agency was not the Federal Government but some private charitable organization such as a church or CARE. The Federal Government acquired title to the food and then gave it to the charitable group.

As prosperity or some measure of economic vitality returned to many countries, Congress decided the recipients should pay for some of the help and save the United States some money. In 1959, Congress amended Public Law 480 to include Title IV, which was first used in 1961. Under this title, the United States loaned money to buy food. The recipient country repaid the long-term loans at very low interest rates. This program, called Food for Peace, often has been considered apart from general farm legislation which underwent changes simultaneously with the development of the Food for Peace program. One can hardly be understood in isolation from the other.

PRICE SUPPORTS

In 1948 and 1949, Congress continued price supports for most farm commodities. The formula called for 90 percent of parity if Congress appropriated the money. For some time critics had attacked the absurdity of maintaining farm income at an equivalent of the so-called best years for U.S. farming—from 1910 to 1914.

The program had become increasingly cumbersome as the parity items changed. Farmers no longer bought horse harness, but they did buy television sets. And who could do more than guess how many bushels of wheat at 1910–1914 prices would buy a television set at a time when there was no television? So Congress revised the parity formula to use a ten-year moving average instead of the fixed averages of 1910–1914. The law also provided for production quotas and restrictions on crops, much as before. These acts corresponded with the end of the Marshall Plan and the fear of a postwar depression which never came.

No general depression materialized, but for farmers nothing seemed to help except war. As the Korean Conflict continued, disputes in Congress and elsewhere went on concerning the issue of flexible rather than fixed price supports. Should price supports be set at a fixed percentage of parity, or should the percentage be changed as circumstances changed? What really was at issue was the overproduction of certain commodities simply because they were supported, and the underproduction of others because they were not supported adequately. A drop in parity percentage in wheat and a rise in percentage for cheese might move farmers out of wheat into dairying. Some felt that such gentle coercion might well serve the national interest. Others supported flexible price supports in order to reduce supports and drive farmers out of the business altogether. Scholars and politicians pointed out the surplus farmers who should be forced into other occupations. Flexible price supports seemed one way to achieve this goal.

The goal was to get rid of federal price supports altogether. If only those who could survive without help stayed in agriculture, then Congress could end all supports; but U.S. farmers needed an export outlet. Many hoped Food for Peace would establish a prosperous world market for American farm products. Simultaneously, Congress sought a world market through generosity and trade agreements, and it encouraged the flight of the less successful entrepreneurs from the farms.

THE SOIL BANK

Ezra Taft Benson (secretary of agriculture under Eisenhower) got his way in 1954, and Congress agreed to flexible price supports. Farmers were ruined by the thousands. To supplement flexible supports, Benson came up with the idea of the Soil Bank in 1956. This device allowed farmers to lease their farmland to the secretary of agriculture who, in turn, would put it in grass or another soil-conserving crop. This idea dated back at least to 1933 and Henry A. Wallace of the New Deal. The Soil Bank as such lasted until 1958, proving in the interval that the program worked no better than it had during the Great Depression.

The Soil Bank did not halt the steady emigration of farmers from farming; in fact, it may have increased this movement. But American farmers became more proficient (Table 26.1). Food became the great weapon of the cold war.

The margin between overabundance and a shortage was always small in terms of total production. However, the comparatively small amounts of food distributed in foreign aid made a considerable difference in the income of farmers. More importantly, the aid programs outside the government apparently stimulated agricultural exports as Congress had hoped because their value increased from 73.5 percent of total exports in 1955 to 81.8 percent in 1969.

The percentage and amount of ordinary trade exports rose consistently as the Food for Peace and related programs declined. This could have been attributed to governmental closing of the programs. In fact, the dollar amounts sold or given away steadily rose until 1965. The trade exports exceeded any decline in the charitable enterprises.

The programs and exports had considerable influence on the total

TABLE 26.1. Number of Farms and Cropland Harvested, 1945–64

Year	Number of Farms	Cropland Harvested* (preceding year)	Av. Cropland Harvested (per farm)	Av. Size of Farm	Sharecroppers	Total Land in Farms
			%	acres	%	acres
1945	5,859,169	352,866,000	60.2	194.8	15.5	1,141,615,000
1950	5,383,437	344,399,000	63.9	197.0	13.1	1,161,419,720
1954	4,782,416	332,870,000	69.6	242.1	11.6	1,158,191,511
1959	3,710,503	311,285,000	84.0	302.7	7.4	1,123,507,574
1964	3,157,857	286,708,000	90.9	351.5	...	1,110,187,000

SOURCE: *Statistical Abstract*, for the listed years.
* Excludes Alaska and Hawaii.

farm economy. In the only commodity series consistently reported—cash grains, which included wheat and rye, but not rice—the percentage of total production exported regularly rose from 19 percent in 1960 to 26 percent in 1967 when Food for Peace was booming.

Rice and cotton remained comparatively high in the export market. Americans exported about half their rice and about 40 percent of their cotton from 1946 to 1970. Other agricultural products also benefitted from rising foreign sales and from the Food for Peace efforts. The several governmental foreign aid plans probably improved life in the recipient countries as well as increasing the food exports of the United States.

CONSUMPTION TRENDS IN THE UNITED STATES

Per capita consumption of a few foods in pounds per person shows nothing startling (Table 26.2). The per capita consumption of fresh fruits and vegetables steadily declined. Frozen vegetable consumption did not make up for the loss. This pattern of eating ran contrary to rules of nutrition and makes one wonder how Americans could keep healthy eating fewer dairy products and fresh fruits and vegetables.

The answer might be found in vitamin pill consumption. In 1945, Americans consumed some 2,516,000 pounds of vitamin pills. This rose to 2,981,000 pounds in 1950 and to 5,131,000 pounds in 1955. Consumption increased to 7,995,000 pounds in 1960 and stood at an unbelievable 12,324,000 pounds of pills by 1968! Vitamin pill consumption rose 386 percent between 1945 and 1968! Most other food increases or decreases changed in small percentage points. Total population came to 150,697,361 in 1950, of which 59 percent was urban; the 1960 population came to 179,323,000, of which 63 percent were urbanites; and the 1970 census showed some 204 million with about 70 percent urban.

Part of U.S. surplus wealth unquestionably resulted from the comparatively low cost of food. This resulted from the efficiencies of

TABLE 26.2. Per Capita Consumption, Civilian

Year	Meat	Potatoes	Wheat Flour	Dairy Products
		lbs per person		
1950	144.6	106	135	477.8
1955	162.8	107	123	472.5
1960	160.8	111	118	436.7
1965	166.7	114	113	411.3
1969	181.1	124	111	376.5

SOURCE: *Statistical Abstract*, relevant years.

the farmers and of the entire market mechanism—from transportation to retailing.

TRANSPORTATION DEVELOPMENTS

As farm population declined, the number of automobiles on farms also steadily declined from 4,140,000 in 1955 to 3,587,000 in 1965. However, total farm-owned transport rose. The number of farm trucks increased slowly from 2,610,000 in 1954 to 3,185,000 in 1970. Farm tractors increased in spite of the reduction in the number of farms. The number of horses and mules on farms steadily decreased. In 1954, the census figures for these animals combined came to 4,141,000. After 1960, even the combined number was not formally reported. The land released from the production of horse and mule feed became available for other farming.

No one knew how much agricultural produce moved by truck. Statisticians estimated the number of trucks used for hauling agricultural products by a sampling process based mostly on categories of vehicle registration. In 1967, the estimated agricultural trucks came to 3,710,000. In comparison, wholesale and retail trade accounted for 1,887,000, construction for 1,444,000, and utilities for 1,231,000.

Within the broad truck classifications, agricultural use dwarfed any other category except the class called "personal" which probably included many farm trucks. No doubt, with the cost of food and the price for farm products both declining, other industries easily could have surpassed the value of farm shipments.

Any discussion of trucking has to include the development of "piggyback" shipments on the railroads. This began seriously only after 1945 with the use of special flatcars designed to hold truck trailers. This type of freight handling by the railroads increased through 1972. Piggybacking worked best for processed foods. Contrary to predictions, the method did not noticeably reduce highway congestion.

To meet the needs of more passenger cars and to keep the trucks rolling, the highway system of the United States was vastly increased. The linear mileage does not really tell the story, for many new highways had multiple lanes; and the older secondary roads were gradually improved, repaired, and often widened. The mileage figures alone show substantial growth, however. In 1950, federally aided highways totalled 641,000 miles. This had risen to 911,000 miles of federally assisted highways by 1968. Total mileage, rural and urban, increased from 3,313,000 miles in 1950 to 3,705,000 in 1967. The movement of farmers and farm products influenced the highway construction programs of the 1950s and 1960s.

Not only did the highways increase in mileage and in lanes, and not only did the number of trucks rise, but technological changes occurred in the automotive and related fuel industries. Engines became more powerful, lighter in weight, and more economical.

CHANGES IN THE HANDLING OF PRODUCTS

The handling of farm products also experienced technological innovation and some savings as a result. Handling savings in cotton warehouses, for example, allowed a reduction of charges in 1959. Hand labor costs in trucking, pushing, and lifting had been eliminated with the use of forklifts and allied equipment. The savings in charges for some items came to as much as 40 percent or more of the charges in 1959. This redounded to the benefit of cotton producers and to consumers as well.

Grain required special handling. To prevent fires, explosions, and spoilage, grain had to be turned. That is, it had to be moved (generally by conveyors) from one bin to another to cool the grain, to allow sampling, and to fumigate for insects. The grain usually had to be turned about four times a year at considerable expense as it had been done since at least 1874. In addition to the power and machinery needed, turning required empty bins for the transfer which could take 20 percent or more of the available storage space. Furthermore, turning caused breakage in the grain, shrinkage, and a consequent reduction in its value. The obvious solution was to move air through the grain, rather than grain through the air.

Farmers had long practiced grain aeration on a small scale where small storage bins made it possible. Ships had sometimes aerated grain on a small scale. The Department of Agriculture and engineers in state experiment stations undertook research about 1954. They had already worked out solutions by 1955. Grain handlers expected to shift 90 percent of their facilities to forced air treatment by 1961, with savings of about 16 billion dollars a year. The savings resulted from cutting the cost of the process, utilization of formerly wasted storage space, and less fumigating against insects. The farmers got some of the savings, but they were diffusely spread. In an inflating economy, savings could be unnoticed statistically.

DECENTRALIZATION OF LIVESTOCK MARKETING

In livestock marketing, the most significant change was its decentralization. This had begun as early as World War I, and by 1950, about two-thirds of the hogs and one-fourth of the cattle went directly to a market instead of a central terminal. The culmination of this trend occurred in 1970 when the Chicago Stockyards closed. An era of labor exploitation, oligopolistic pricing, and general inefficiency came to an end.

Farmers had to send their livestock somewhere, and local packers came into business or expanded their activities. Many were inadequately inspected or regulated, often because they confined their business within their state borders. In 1965, Congress passed new rules controlling these packers and brought some order and sanitation to the business. Some of the packers fell to the status of conglomerate parts, and many of them experienced serious economic troubles which centered mostly in the retailing business.

Possibly the decline of the central market resulted from the use of the truck as much as anything. The central market made some sense when cattle and meat both moved primarily by rail. With trucks, farmers could move their livestock short distances to market. Processors readily redistributed the meat to the stores and to the consumers by truck. In some places, local packing plants bought from the surrounding farmers who hauled their cattle to the plant and took what the meat processor would pay. The price quoted usually bore some relation to the prices given at a large terminal such as Chicago. In the 1950s, the development of a futures market and trading, as in wheat or cotton, provided a central pricing system without an actual market for physical transfer of the animals. So the system of direct buying could flourish with a minimum of distrust and cheating.

The other method of marketing was auction selling and buying. By 1951, some 2,500 auctions operated in various parts of the country. Farmers brought their stock, and several packers bid for the animals. Many of the so-called packers were actually retailing chains such as Safeway. The auction markets had fair communications with the outside world and central commodity markets. The system minimized distrust and inhibited excessively low bids. At the same time, livestock feeders increasingly bought directly from ranchers. The annual trip West by farmers in Illinois, Iowa, and other places was not uncommon.

In the late 1960s and early 1970s, various reformers again demanded increased consumer protection from fraud and filth. Most conspicuous was Ralph Nader who began with an assault on the automobile industry. With his corps of assistants called "Nader's Raiders," he next began to check into food and food processing. They seemed destined to have more than passing effect. The change in consumer power flowed from changes in the processing industry. In part, changes in meat processing resulted from changes in the retailing business. Following World War II, the great retailers had gradually moved into the position of primary power. The retailing chains slowly became the masters of the processors.

DOMINANCE OF SUPERMARKETS

The dominance of the supermarkets could have been predicted. One supermarket led to another, and before long a chain appeared. As late as 1960, however, the large number of small food stores surprised some authorities. The efficiency of retailers did not begin to match the greater efficiency of American farmers. The supermarket dominated the food retailing business and the food buying business. By 1957, supermarket chains did about 20 percent of all the retailing and steadily increased their share from 45.7 percent in 1960 to 52.4 percent in 1969.

Farmers complained that this selling power enabled the supermarket chains to dominate the livestock market and often force concessions from the processors. Some supermarket chains not only had their own brands of food (canned and frozen) but also bought feeder cattle and finished them on company farms. If the store chain overproduced, however, then it was in the same pickle as any other farmer. Nevertheless, feeding on contract, or vertical integration, could be profitably undertaken by the great chains.

Official and private charity and domestic and overseas trade in food products assured the farmer of outlets for his food. He produced ever vaster quantities with greater efficiency. What could not be sold could be stored. Federal policies cut back production some. However, price indexes for all farm commodities revealed problems (Table 26.3).

So, in spite of all the advances in marketing, processing, and retailing, and generous donations abroad—the American farmer still kept producing himself into a hole. Feeding the hungry of the world was terribly expensive. But changes in all markets seemed to be under way as the 1960s closed.

TABLE 26.3.   Farm Price Indexes, 1945–69

Year	Farm Price Index Average (1910–14 average = 100)	Parity Ratio, 1910–14 (Presumed fair price = 100)
1945	207	109
1950	258	101
1955	232	84
1960	238	80
1965	248	77
1969	275	74

SOURCE: *Agricultural Statistics*, relevant years.

In 1968, Americans "discovered" hungry people in the United States. Feeding Americans cost less than feeding hungry people elsewhere. And feeding Americans would yield good returns to the American farmer. The welfare programs needed revamping so that the system could more efficiently supply the poor with large amounts of food. Politicians and humanitarians also had to define the poor in terms that classified more people as poor, and thus potential consumers of farm surpluses. The machinery of the food stamp program, a device for getting food to the poor, had to be made more efficient. Riots furthered the idea of a domestic Food for Peace program.

Farmers had nagging doubts about the whole idea of crop restrictions. Let the land lie idle while men starved? The reformers again offered a way out. A well-fed generation of rebels put ecology above economy. They insisted on saving the land for open places and halting the advances of technology. Science and technology enabled the continuing production of surpluses. Farmers could benefit from this new turn of events. Being ever smaller in number and better able to organize, they might simultaneously cut production, conserve the environment, and feed the hungry of America!

CHAPTER TWENTY-SEVEN

THE TECHNOLOGICAL CHASM ⚜ 1945–1972

T HE UNITED STATES has always been one wing of western civilization. The Republic had been founded on the highest principles of the Enlightenment. No serious break had ever cut it off from contact with Europe. Building on its European heritage, the United States had become Europe's finest achievement. Americans had taken a European pattern and made it work. The United States eventually pulled far ahead in technological skills and in the material comforts these brought. A technological gap between America and the rest of the world was widening in the 1960s. Some even proclaimed the existence of an unbridgeable chasm.

One argument ran that high-quality public education was a peculiarly American achievement. This reservoir of skilled talent kept making more discoveries, and the educational level of the American farmer allowed him to make effective use of the discoveries. Farmers also could afford to adopt new and expensive methods and devices which, in turn, increased production and income.

By the end of World War II, several varieties of plows were available for different soils and climates. Farmers increased efficiency by using more plow bottoms which required larger, more powerful, more economical tractors. Tractor attachments, power takeoffs, and hydraulic lifting equipment were in use. Five- and seven-bottom plows became commonplace, and the tractors could pull these monsters rapidly through almost any terrain.

Rapid plowing had to be accompanied by equally fast seeding. In the postwar years, the single-row planter gave way to six- and eight-row drills of greater size and efficiency. Using a multiple-row drill enabled a farmer to plant 80 or more acres a day as compared to 7 acres with a one-row planter. The drill had to suit the specific conditions of the soil, the climate, and the plants. The shift to more plow bottoms and gang-operated drills was expensive, but it impressively increased farmer efficiency and income. The steady migration of farmers off the land gave those who remained higher acreages to farm, although the total farmland in crops fell significantly nationwide.

Research and experimentation, much of it government-financed, produced seeding equipment to meet varying conditions. Farmers often seeded and fertilized at the same time, particularly grass and legumes. The machine laid a strip of fertilizer and put the seeds directly on top of it. Other drills placed seeds on ridges or beds with furrows in between for drainage or irrigation.

Where wind erosion created problems, seed drills were designed to place seeds in deep furrows. This method left less land broken and open to erosion. Deep drills had special soil openers and farmers usually ran them in gangs. Simple flat seeding was common on the prairie, especially in the Midwest. Wheat, rye, and corn did not do as well if planted on top of fertilizer. Special drills that placed the fertilizer to either side of the drilled grain came on the market in the 1960s.

Planting machinery had furrow openers that were usually a variation on duck-footed or shovel plows. Such openers served well for simple seeders, but they required more draft for multiple drills and did not work well at higher speeds. Two improvements were made in the 1950s and 1960s. First, the shovel was reduced in size and a shield put on it to hold the furrow open with a decrease in resistance. Second, double disk openers were used. Some farmers even drilled seeds as they plowed. The drills planted in the tracks made by the tractor wheels, thus leaving loose soil on the sides.

Plowing and drilling in gangs and applying one dressing of fertilizer at the same time demanded powerful tractors, but two or three operations were done at the same time. The precision planting produced better stands of crops—stronger and more uniform. The use of drills also reduced seed and fertilizer costs.

Depending on the region, farmers seeded hay and grass in spring or fall. Some plants were usually seeded at one specific season, such as red clover in the spring. Typically, several grasses and legumes were grown in the same field. The newly seeded field was rolled. The logs of earlier days had long since been replaced by corrugated iron rollers. In the mid-1940s, some farmers still broadcast grass seed, but drilling became much commoner.

In the postwar years, the Department of Agriculture and other agencies pushed grassland farming in order to halt wind and water erosion and to reduce the acreage in surplus commodities, such as wheat, cotton, and corn. Although the herbicide 2,4-D was discovered during the war, grassland farmers had to rely on mowing weeds as a control measure until well into the 1950s. Even then, farmers had to mow fields of legumes. For the most part (seeders excepted), farmers could use their usual machinery for grassland farming.

HARVESTING MACHINERY

After 1946, most wheat was harvested by combines. Some crops like soybeans were windrowed first and picked up and separated by combine later. Tractors pulled most combines as late as 1960. The combines operated off a power takeoff and cut comparatively narrow swaths 6 to 12 feet wide. With a reasonably small acreage, a tractor-pulled combine could do the job. As acreages increased, the self-propelled combine came into use, but it accounted for only 20 percent of combines in 1960. The self-propelled machine had obvious advantages. These combines had larger cutter bars with various devices for altering height of cut, speed, and other things. In 1954, attachments were available which made the combine usable for either small grains or corn. The farmer also had better visability. Other conveniences included comfortable seats, radios, and air-conditioning.

Automation appeared early in self-propelled equipment. One example was the automatic shift. Another, in the combine, was automatic leveling of the separator, regardless of the pitch of the ground and the rest of the machine. The mechanism depended on electric switches to start the leveling machinery which used hydraulic pressure to do the work. Self-leveling may have seemed less than revolutionary in a technological sense, but it seemed a minor miracle economically.

The field shelling of corn was an important development. The corn attachment on a combine made shelling in the field possible. The multipurpose combine speeded corn harvesting. The portable corn drier appeared around 1949, probably in response to the need created by the first picker-shellers. Farmers could not use field shellers if their grain had a high moisture content, and they had to use expensive, cumbersome driers. But portable drying equipment was soon available when field shelling became possible. Beginning in 1958, standard corn pickers could also have shelling attachments added. Manufacturers made some 4,683 of these picker-shellers in 1959. Production fell to 712 in 1960 and rose to 1,127 in 1961. Thereafter picker-shellers disappeared from the records in *Agricultural Statistics*.

The production of corn pickers tended downward from 34,811 in 1959 to 13,797 in 1968. Simultaneously, corn-picking units with combine attachments rose from 5,324 in 1959 to 25,576 in 1965. However, only 16,655 were produced in 1968. Such declines in the farm equipment business reflected both the decreasing number of farmers and their decreased buying power. Use of the combine-corn picker-sheller with field shelling of corn was the chief method of corn harvesting by 1965. Some farmers still stored corn on the cob and shelled it in stationary machines after the corn had dried.

During World War II, the successful spindle cotton picker and the tractor-mounted cotton stripper had appeared. The big migration of blacks to cities came after the development of the cotton picker. Mechanization of the harvest was almost complete by 1970. The development of herbicides, defoliants, and easily picked cotton varieties all attracted attention as soon as the spindle cotton picker proved successful. Mechanization in cotton husbandry substantially reduced the time required in growing and harvesting an acre of cotton. The gap between mechanical pickers and tractors on one hand and mules and handpickers on the other came to a ratio of 6:1.

INCREASED USE OF SILAGE

Greater efficiencies in feeding resulted from the expanded use of silage. After 1950, new improvements in silage machinery brought savings in time and in energy because the stock raiser could handle greater volumes of feed. Tower silos increased in number. From 1945 to 1972, the principal silage crop remained corn, although other plants were ensiled. Corn made up about three-fourths of the silage. Sorghum declined in ensilage use.

Two breakthroughs had been the field forage harvester and chopper and the silo unloader. Both machines increased in number in the postwar years. By 1960, farmers cut and chopped over 90 percent of all their silage in the field. Methods of loading the silage into the silo varied: some farmers blew it into the silo, some used elevators, and some used endless screws. The blower used a large amount of power and did not make up for the added cost in speed, compared to other methods. The elevator, run from the tractor power takeoff, did the job cheaply and effectively and became the preferred method in the 1970s. The farmer still had to regulate the flow of silage from the wagon to the blower or elevator.

IMPROVEMENTS IN HAYING

The lack of any easily obtained statistics on silage and the abundance of statistics on hay (and the vaguely labeled "forage") suggested that silage accounted for far less of the fodder crops than did hay, forage, or feed grains. Hay remained the most common feed. Haying traditionally had caused farmers the next most work to plowing. Even after the mower came into use, the hay crop required raking and unloading.

In the 1950s, the tractor finally replaced the horse as power for the mower, and the mower bar was lengthened to seven feet from the

Harvester, 1964 model, in Kansas wheat. Courtesy USDA

previous five feet. With a tractor 30 to 50 or more acres could be cut in a normal day. The mower laid the grass in windrows. Manufacturers incorporated many safety devices such as a safety spring release that threw the cutter bar back when it hit an obstruction. Tractors usually pulled mowers, using power from the power take-off. Operators could also mount mowers at the rear or side of the tractor. Except on large-scale farms, tractor-drawn mowers were commoner. High-speed cutter bars characterized the mower from 1950 to 1970.

The field pickup baler was in use around 1932, but it still required men to tie the bales. In 1940, self-tying balers came on the market. Tractor-pulled, they speeded the work and reduced the drudgery. The self-propelled hay baler appeared in 1958. By that time, twine was commonplace for binding. Of the 600,000 balers in the United States in 1958, around 480,000 used twine rather than wire. The number of pickup balers on farms rose from 42,000 in 1945 to 795,000 in 1965.

One man could handle small bales, and so small, rectangular bales were commonest through the 1960s. Automatic bale ejectors and conveyor belts carried the bales into the hayloft. As in other farm machines—clutches, flywheels, and other safety devices became standard to protect both the machines and the operator. Nearly foolproof tying devices had become standard during the 1960s.

FERTILIZER APPLICATION

Soil fertility had to be increased or maintained. Fertilizer application varied widely. The farmer applied it at different stages, determined by the crop and by the kind of fertilizer. Manure, rich in organic matter and plant nutrients, was spread over the field from the tailgate of a manure spreader and plowed in before seeding. The spreader had a regulating mechanism, consisting of agitators shaking over an opening which spread manure over a few rows behind the machine. By the 1950s, fertilizer trucks had come into use on many farms. Manure spreaders on trucks had the advantage of great capacity, sometimes 8 or 10 tons per load. Such trucks used complicated, effective spreading and metering machinery involving tubes and augers for dry manures and fertilizers.

In the 1950s, a new metering device for the pulled spreader appeared that consisted of a notched disk over each outlet, replacing the rotating bar wheel of the earlier machines. Accurate measuring of manure assured uniform application. These machines were usually used before plowing.

Farmers frequently applied dry fertilizers and unpressurized liquids in bands under or to the side of the seed at the time of seeding. Drills now had fertilizer attachments, some integral and some attached. Plows of every description from subsoilers to chisels could have applicators for fertilizers attached. In any case, the fertilizer was measured and applied during or very close to the seeding operation of grains, grasses, and vegetables. The application was controlled by devices run off the power takeoff or sometimes off ground wheels, depending on the machine and the lay of the land. To spread fertilizer uniformly on uneven ground or at irregular speeds, a metering device depending on the ground speed rather than engine speed did the best job.

Fertilizing of crops after planting was called either top-dressing or side-dressing. Meadows and pastures could be top-dressed simply by broadcasting, but grain and vegetable crops usually required side-dressing. Application was done with either harrowing or cultivating, and the harrows and cultivators had fertilizer applicators attached. If the farmer applied liquid fertilizer, he could put it on along with his sprays for insects and weeds.

Farmers used all sorts of pumps and meters for liquids not under pressure. The most satisfactory measured and applied according to the forward speed of the applicator. Otherwise, the pumps kept

Corn harvesting, Iowa, 1973. Photo by Charles Benn

going if the tractor stood still, or they did not deliver enough when it moved forward too fast. Whether for dry or liquid fertilizer, those applicators used in conjunction with harrowing or cultivating usually had shovels or disks to open the soil in front of the nozzle or opening.

ANHYDROUS AMMONIA

The big news in fertilizers was the discovery and ever-widening use of anhydrous ammonia soon after World War II. This pressurized gas turned into a liquid when released into the atmosphere. As late as 1959, some 37 percent of farmers applied no fertilizer to their commercial crops. Anhydrous ammonia changed this figure during the next decade.

Along with everything else, fertilizers rose in cost, and more careful use of seed and fertilizer became necessary. So farmers minimized fertilizer use during the uncertain years of the late 1940s and after the Korean Conflict. Anhydrous ammonia offered cheap, effective fertilizing, although supplemental fertilizers were required. Machines to apply it efficiently appeared quickly. The high-pressure fertilizers required some extra capital expense, but they took little labor to apply and cost little in the long run. Farmers could apply anhydrous am-

Silos, Iowa, 1970s. Photo by Charles Benn

monia almost anytime until harvest. Orchards and vineyards could be fertilized until fairly late in the season.

EQUIPMENT FOR ANIMAL HUSBANDRY

Animal husbandry may not seem to involve much application of mechanical and technological devices. But in dairying, harvesting the animal product involved machinery. Milking machines became increasingly popular during World War II, and bulk handling of milk seemed logical. The necessary tanks and machinery were invented during the war.

Bulk handling of milk—moving the product by pipeline from milking machine to storage tank and by another line to tank truck—became important commercially in the years from 1948 to 1950. Dairy processors and creameries began offering premium prices to dairy farmers who would install the necessary machinery. At the same time processors began to penalize farmers who did not use the new machinery. By the mid-1950s, a dairyman found it virtually impossible to have his milk picked up if he delivered it in cans. By the 1960s, hardly any commercially produced milk was handled in cans. Milk cans became items for antique collectors. Bulk handling of milk on the farm ultimately produced tremendous savings. Although some of the savings went to farmers, most went to the dairy processors.

Machinery for delivering feed to mangers and for cleaning out manure had been invented earlier. Advances in animal productivity through breeding improvements and better care resulted in greater per-animal consumption of feeds. This trend continued until the late 1950s and 1960s when the discovery and use of various feed supplements increased feeding efficiency. Until then, animals used increasing amounts of feeds, which were available and which were needed in larger quantities by the improved animals. The farmer had an obvious incentive for more efficient feed distribution beyond being able to handle more animals.

Probably one of the most important changes in mechanical feed operations occurred in the handling of silage. The stock farmer usually had to do the job twice a day. Filling the silo had been done efficiently for some time, but a man with a fork or a shovel had to put it in the chute that took the silage down. Unloading from the bottom presented the same difficulties. In the 1950s, electrically powered silo unloaders appeared. From 1955 to 1958, some 13,000 silo unloaders went into operation on U.S. farms, and this number increased in the 1960s and 1970s.

American farmers used two types of unloaders. The top unloaders had conveyor belts from the top of the silos which ran on large wheels on the surface of the silage and then up and out. The bottom unloaders worked chiefly as augers. Both types could deliver around 150 pounds of silage a minute, frozen or unfrozen, by the 1960s. The silage business had been completely mechanized.

Machinery got bigger, and in the United States, bigger meant better. The 1943 spindle cotton picker placed on exhibit in the Smithsonian Institution in 1970 was so large it barely could be moved into the museum. It weighed 9,000 pounds, and it picked only one row. The first of its kind, it was affectionately known as "Old Red" and was regarded by cotton men as a quaint little antique.

The maintenance of equipment was very important. The farmer kept his array of machinery in operation mostly by his own efforts. He had to be an expert mechanic. The farmer made trips to town for replacement parts, but he often could not wait to get parts. By 1970, it was a poor farmer, indeed, who did not have an electric welding outfit or at least a neighbor who had one. A farmer who could not afford the tools to maintain his equipment could not long remain in farming. He had to understand internal combustion engines, electric motors, and hydraulics as well.

The technological knowledge also demanded a comparatively high level of education as well as of practical skill. If the farmer could not understand the owner's manual he was lost. Americans had invested heavily in agricultural education with the land grant colleges, and the rewards in food production had been enormous.

CHAPTER TWENTY-EIGHT

BACKWARD, TURN BACKWARD ⚓ 1945–1972

ALL PEOPLES, primitive or sophisticated, have memories of at least one Golden Age when men were Virtuous and Happy. The tendency to praise every age except one's own and every country except one's own seems a universal human passion. The problem apparently arises from an inability to understand the present.

In 1937, Paul Sears wrote a book, *Deserts on the March,* wherein he predicted the destruction of the United States as deserts engulfed its farmland. Thirty years later, that primal destruction of soil erosion which formed the base for the word "pollution," that is, "to make dirty," had been stemmed to a remarkable extent. Then in the late 1960s and the early 1970s, alarmists "discovered" chemical pollution, which they proposed to end by turning back to a Golden Age without DDT, 2,4-D, and other potentially harmful biochemicals. The alarm had been sounded loudest by Rachel Carson in *Silent Spring* published in 1962.

But the golden age of biochemistry had just dawned. Before the discovery of the double helix of DNA in 1953 by James D. Watson and Francis Crick ushered in the age of molecular genetics, breeding might have seemed incongruous under the heading of biochemistry. But now genetics fell under its rubric. How did it happen that a simple hybrid—a cross between a Brahman bull and a red Shorthorn—produced a bull calf whose offspring bred true? The answer lay hidden in the double helix, but the reality of the cattle was in Texas.

In 1920, the King Ranch, by genetic chance, produced the bull named Monkey from which Santa Gertrudis cattle were developed by careful breeding. Monkey was bred to Shorthorns, and his progeny bred and rebred until the King Ranch had a sizeable herd that bred true. By 1932, when Monkey died, the breed had been established on the King Ranch.

In 1940, the Bureau of Animal Industry recognized the Santa Gertrudis as a true breed. The new thick-skinned cattle resisted insect pests and survived high temperatures and water shortages. The King Ranch made its first public sale in 1950, and the breeders of the Santa Gertrudis formed their own association in 1951. The astonishingly productive breed soon spread throughout the South.

The success of the Santa Gertrudis led to a rash of new breeds which took time because development depended on genetic good luck. The Beefmaster, a true-breeding cross of Brahma-Hereford-Shorthorn, appeared in 1956. The Brangus and the McCan, combining Brahma genes with Angus and Herefords respectively, came about 1958. Beef production increased as a result, but feed and care still counted most. Begun by dairymen, artificial insemination now contributed to a rapid dissemination and increase of the new breeds of beef cattle.

Dairy herds consistently improved milk production. Department of Agriculture statistics showed the superiority of progeny of artificially bred cows. Purebred cows averaged 10,580 pounds of milk and 430 pounds of butterfat in 1947. This rose to 11,304 pounds of milk and 477 pounds of butterfat by 1957. The number of cows rose from 1,184,000 in 1947 to 6,056,000 in 1957. In comparison, the average nonpurebred cow produced only 5,007 pounds of milk and 199 pounds of butterfat in 1947. This rose to only 6,160 pounds of milk and 235 pounds of butterfat in 1957.

Similar developments in hogs and poultry resulted in animals that could take maximum advantage of the new advances in feeding and handling. In 1946, the Minnesota Experiment Station bred a new type of meat hog called Minnesota No. 1, first in a line of improved hogs. That same year, another new type of meat hog called the Hamprace also appeared.

Between 1945 and 1970, beef production rose from 10,275 million pounds in 1945 to 21,158 million pounds in 1969; pork went from 10,697 million pounds in 1945 to 12,953 million pounds in 1969. The number of chickens increased about ten times.

ADVANCES IN ANIMAL NUTRITION AND ADDITIVES

Progress in animal nutrition resulted in stronger and more productive animals. The first organic chemical to be synthesized was urea. By the 1930s, its value in feeding was well known. However, it was not an important source of protein as long as cottonseed and soybean meal remained abundant. The war created a shortage of feeds, and meat prices went up. Feed shortages and high prices encouraged animal raisers to supplement their feeds with urea. Commercial production of urea became profitable and reduced urea costs after the war. Urea became increasingly important in livestock and poultry feeding.

Stock farm, North Dakota, 1940s. Courtesy Library of Congress

The business of manufacturing feeds began to boom. In 1947, farmers used 119.3 million tons of concentrated feeds, of which 26.0 million tons had been manufactured, or 21.8 percent of the total. Manufactured feeds amounted to 22.8 percent of the total in 1949. By 1950, the total of concentrated feeds had risen to 125.9 million tons, and the manufactured share had gone up to 23.1 percent of the total. Most manufactured feeds were supplemented with additional protein and vitamins. The value of folic acid for poultry and animal nutrition was discovered in 1946. The chief source of added proteins had been meat scraps and fish meal. Chickens needed about a fourth of their protein from animal sources. Other animals needed less meat proteins, but they did better when fed them.

This raised the question of what the meat contained that other feeds did not. In 1948, Karl Folkers and his associates discovered vitamin B_{12} which turned out to be the unique growth factor in animal protein. In 1949, vitamin B_{12} was offered commercially to feeders and poultry raisers. By 1950, swine and poultry raisers had begun using feeds fortified with B_{12}.

Advances in veterinary medicine became associated with feeds and feeding. The drive to produce antibiotics in larger quantities resulted in the discovery that B_{12} was a by-product of the production of streptomycin and other antibiotics. In 1949, E. L. R. Stokstad and T. H. Jukes found a residue of antibiotic remained in some vitamins so produced, and the antibiotic seemed to provide an additional growth stimulant. Feed processors began adding different antibiotics in small amounts to feeds after 1950.

In 1949, the growth-promoting effects of arsenic also attracted some attention. For some reason, only rats suffered deadly effects from moderate amounts of arsenic. The bodies of most animals did not store it, so it rapidly cleared the system. Most animals benefitted from small amounts, and it did in fact promote growth. However, it did not enter into commercial feed production largely because of popular prejudice against it.

Antibiotics had been shown to stimulate animal growth and to increase their survival rate as well. Scientists soon discovered that the several additives in combination seemed to have effects beyond what might be expected from addition of the single elements. Manufacturers of feeds added synthetic amino acids for protein to feeds in 1951 and to animal fats in 1952. They began to add antioxidants to their feeds to help preserve the fat-soluble vitamins and the fats in 1953. The antibiotics also helped to retard spoilage. Within a decade the animal and poultry feeding industry had been almost entirely transformed.

Because these additives could not be produced on the farm, feeding began to be a function of biochemical factories as well as a business of farmers. *Agricultural Statistics,* published by the Department of Agriculture, did not tell much about these feed additives, their quantities, or the prevalence of their use in 1970. The statistics did reveal that most farmers got their major income from nonfarm jobs. The role of additives and biochemistry in this development went unnoticed. Did the farmer need to moonlight or did he simply have an opportunity to increase his income by holding more than one job?

The new discoveries allowed the farmer to substitute factory feeds for other more expensive feeds and to get better returns at the same time. Part of this change resulted from the discovery of the feeding effectiveness of stilbestrol. In 1951, Wise Burroughs and others discovered that lambs fed on certain hormone-rich feeds gained weight much faster than expected. The discovery, made almost by accident, led scientists at Iowa State University to investigate the effects of the synthetic hormone diethylstilbestrol (stilbestrol) on growth. They found in every case animals gained more while eating less. Additives that made animals hungrier had been discovered, but stilbestrol actually reduced their feed needs in proportion to the growth that resulted.

The farmer could go two ways: (1) use less feed or (2) feed the same amount to more animals. For the most part farmers chose the latter course. The addition of 200 pounds of B_{12} to 15 million tons of feed in 1951 replaced all of the B_{12} naturally occurring in 1 billion pounds of meat and fish. Feed processors substituted urea for meat and soybean meal at a ratio of about one to five. One pound of urea provided the protein supplied by over five pounds of soybean meal.

All of this happened quickly and with little publicity. As early as 1951, farmers and stockmen put $17.5 million worth of antibiotics in their feeds to increase growth and to fight diseases. By 1961, farmers and stockmen used $43 million worth of antibiotics in feeds. At this time, the Food and Drug Administration seemed to have no serious misgivings about antibiotics, but it delayed approval of the use of stilbestrol in animal feeds until 1954. By 1954, between 80 and 85 percent of all beef cattle were on stilbestrol.

As early as 1962, stilbestrol had already increased beef production 10 percent, or by about 100,000 tons. By 1965, feed supplements saved poultrymen enough feed in a single year to have fed an additional 90 million broilers! Because stilbestrol might cause cancer in humans, its use as a feed additive was prohibited late in 1972. Severe meat shortages quickly ensued.

Possibly a 50 percent reduction in premature cattle deaths resulted between 1945 and 1972. The prevention of losses of hogs may have been about the same. For animals in larger herds or birds in large flocks, mass treatment of disease was most effective. Nothing worked quite so easily as adding medicine to feed, but the farmer needed knowledge and skill to use the right medicines at the best time.

FIGHTING FOOT-AND-MOUTH DISEASE AND OTHER DISEASES

Government action in veterinary medicine had long been common. The effects of such intervention ranged from the eradication of pleuropneumonia in 1892 to the virtual end of bovine tuberculosis in 1940. In the 1940s, the greatest remaining threat seemed to be foot-and-mouth disease which posed a constant threat of invasion from the vast reservoir of the disease in Latin America. The disease had no effective cure. The major question facing the U.S. farmer was how it could be prevented.

The traditional method had been to destroy infected animals and to burn any residue such as stalls. In 1947, the disease entered the United States from Mexico. The attack against the disease took its usual form, but cattlemen and others thought the disease should also be fought in its place of origin. So the Republic of Mexico and the United States joined in an eradication program. Some hard feelings resulted because the Americans wanted to eradicate the disease by slaughter, but the wide occurrence of the disease and the fact that many animals were work animals led the Mexicans to insist on vaccination instead. Finally, the Mexicans won, and the United States protected itself with quarantine measures as well as the usual slaughter program. In 1948, Congress established a laboratory to investigate the cause and cure of the disease.

The Department of Agriculture found a new diagnostic test for brucellosis in 1949. The new milk ring tests rapidly identified the animals which required closer examination. The old agglutination test still provided the final, sometimes unreliable diagnosis. In spite of an effective vaccine, the B-19 Strain, slaughter was still the usual approach to brucellosis eradication.

The Department of Agriculture devoted its 1956 yearbook to the recent advances in animal health. The scientists listed an amazing variety of diseases, some unknown but ten or twenty years before. On-farm remedies and professional possibilities appeared in this valuable handbook. Viral diseases still baffled veterinarians and farmers who simply tried to keep the animal alive until it had resisted the invader. Sanitation proved the most effective defense against viral diseases.

For diseases with insect vectors, killing the insects attracted a great deal of attention for both plants and animals. Some insects not only carried diseases but did direct damage themselves. The life cycle of the heel fly included a period when the grub lived inside the host animal, then chewed its way through the hide, and dropped to the earth to renew the cycle. Ticks, lice, and similar pests carried disease organisms and caused sores and other dangerous injuries.

USE OF INSECTICIDES FOR DISEASE CONTROL

Spraying with insecticides had become common by 1945, and a host of new insecticides followed DDT. Chlordane, toxaphene, gamma BHC, aldrin, and scores of others came on the market for use on animals and plants. By 1946, however, scientists proved the harmful residual effects of DDT when it entered the milk of sprayed cows or feed crops that had been sprayed. About the same time scientists discovered that penicillin also entered milk in unacceptable quantities. Milk from cows treated with either DDT or penicillin could not be sold to humans until the animals registered clean. State and later federal laws were enforced.

OTHER METHODS TO ATTACK INSECTS

Another difficulty was the development of immunity to the insecticides by insects. Thus farmers had to use combinations of insecticides in sprays. In addition to the hazards to wildlife and humans, the wide spectrum of the insecticides endangered many useful insects such as bees. In the 1950s, two lines of attack were undertaken: one to find specific insecticides including natural enemies and another to eradicate the insects altogether. The idea of using specific insecticides, natural diseases, or predators for specific species had been tried before, sometimes successfully. The use of the milky white disease of the Japanese beetle eventually brought that bug under control in the 1940s. This method was first used by C. V. Riley to control the cottony-cushion scale with the ladybird beetle in the citrus industry during the 1890s. Finding specific poisons proved less successful.

In the 1950s, two successful methods of eradicating insects altogether were tried. In 1955, U.S. Department of Agriculture scientists eliminated screwworms on the island of Curacao. They used radioactive material to sterilize male flies in huge numbers. The released flies competed successfully with fertile male flies so the females laid infertile eggs. Since the females mated only once, an infertile insemination resulted in no progeny. Done several years in succession, the flies naturally died out, and the insular condition prevented reinfestation.

Subsequently, federal and state entomologists effectively attacked screwworms and other insects on the continent. The workers usually began in Florida and gradually expanded into larger areas. Sometimes the entomologists had to repeat the attack when reinfestation occurred, but when effective, the sterilization of males worked astonishingly well. Farmers benefitted, but a program of this magnitude had to be carried out with state or national resources.

As biological knowledge and cunning increased, scientists developed another method which offered great possibilities in the 1970s. Scientists found ways of synthesizing the several female sex attractants of insects. The use of the sex attractants had been employed for some time on a laboratory scale. Collecting enough females had always been difficult, and breeding them took too much time with meager results. The method of entrapping the males could proceed rapidly only when the necessary attractants could be produced on a large scale. The possibility of luring and destroying all males remained an attractive idea, but it proved less effective than simply sterilizing the males and turning them loose. This approach seemed to offer the greatest prospects for success, but no one had tried it on any large scale by 1972.

The effective eradication program proved attractive to farmers whose losses to insects had been high. These attacks against insects were paid out of public monies instead of farm operating expenses. Spraying cost the farmer his own money directly and had to be repeated. Eradication once done was supposed to be done for good.

Various internal parasites afflicted both animals and their owners. Although systemic poisons had been used for some time, most advances occurred during the 1940s and later. Metallic arsenates—especially those of cobalt, copper, and iron—worked well in killing tapeworms and stomach worms in ruminants. Diethylcarbamazine appeared as an effective killer of the filariasis worm in sheep, goats, and horses. These discoveries took place in 1947. A host of poisons that killed the parasites without damage to the livestock came in the 1950s and 1960s. Although the poisons were expensive, the farmers could administer them in herd and flock treatments.

INCREASING PRODUCTION RATES

The survival and production rates of all animals rose steadily, but unfortunately no one could make a statistical separation of the various causes for improvement. Better animal health unquestionably accounted for part of the increase in meat production and the decrease in prices in some categories such as poultry. In dairying, the average annual yield of milk from a dairy cow rose spectacularly—from 4,315 pounds in 1945 to 9,158 pounds in 1969!

Similar dramatic developments resulted from genetic improvements in plant varieties, particularly the new wheats. Of these, Pawnee, introduced in 1942, seemed most impressive. It resisted insects fairly well and yielded extravagantly. The *Nebraska Farmer* estimated that Pawnee yielded from 25 to 30 percent more than other wheats and that if all the Nebraska wheatland had been in Pawnee in 1946, the crop would have been 8 or 10 million bushels larger. By 1946, Pawnee had become the predominant variety in most of Kansas. By 1950, varieties of Pawnee, Comanche, Wichita, and Triumph occupied 76 percent of Kansas wheat acreage, and more improved varieties were developed during the next 20 years. The wheat yields continued to increase.

As with animals, the new varieties made other improvements possible. Foreign scientists had discovered that applications of fertilizer had important effects on crops only when the plants genetically could make maximum use of the fertilizer. Suitable crops, fertilizer, and trained farmers were needed to close the gap between the United States and the rest of the world. Financing had to be provided, and then the foreign consumer had to be induced to want the new commodity.

TABLE 28.1.   Fertilizer Used on American Farms, 1945–65

Year	Nitrogen	Phosphorus	Potassium	Total
		thou tons		
1945	634	614	539	1,787
1955	1,961	998	1,548	4,507
1965	4,581	1,537	2,356	8,474

SOURCE: USDA, ERS, *Statistical Bulletin 233*, 1966.

ANHYDROUS AMMONIA FERTILIZER

The use of anhydrous ammonia as a fertilizer could be considered the most significant advance from 1945 to 1972. This nitrogen-type fertilizer was stored as a liquid gas under pressure. Equipment to deliver the gas was developed in 1947. The gas entered the ground through nozzles where it became a liquid fertilizer in the ground. The method of storage, transport, and application was cheap and easy with the proper equipment. Farmers could distribute it uniformly in the soil in exact amounts. In combination with other fertilizers, anhydrous ammonia and other forms of nitrogen could substantially increase yields and reduce soil fertility depletion. The statistics on increased fertilizer use, particularly nitrogen (Table 28.1), showed parallel figures for increased production and reduction of crop failures.

In spite of governmental efforts to reduce production in many crops, the trend was substantially upward in most cases. Only vegetables and tobacco seemed to fluctuate. Carefully controlled experiments could assign the exact amount of increase in plant production which scientists could attribute to plant varieties, cultivation, fertilizer, irrigation, good weather, and so on. In the field, the proportion of the higher yields which resulted from new varieties or from culture practices remained a matter of speculation.

Every farmer knew he fertilized his weeds as well as his crop. In dry years and dry regions weeds especially cut returns by using precious water for their own useless growth. Weeds were objectionable for many reasons, and they had been fought with various techniques for centuries. In the 1940s, systemic hormone herbicides of astounding effectiveness and selectivity appeared, and their use has greatly increased.

PROBLEMS WITH 2,4-D USE

The herbicide 2,4-D came on the market in 1945 for public testing. In one year production increased nearly 500 percent, and the quantity sold rose from 631,000 pounds in 1945 to 5,315,000 pounds in 1946! By 1950, farmers applied the new herbicide to 30 million acres of cropland in the United States, and total production exceeded 14 million pounds.

In the meantime, manufacturers made inexpensive sprayers, and airplane spraying became increasingly important. From one-half to one pound an acre could do the job, although farmers sometimes used as much as three pounds an acre.

In 1948, more sophisticated machinery came on the market to distribute herbicides, and volumes as low as three gallons an acre could do the job. Production of 2,4-D was 36 million pounds in 1960, 54 million pounds in 1964, and 79,263,000 pounds in 1968. Another herbicide was 2,4,5-T, of which 11,434,000 pounds were made in 1964 and 17,530,000 pounds in 1968.

As the volume increased, the cost per pound tended to fall significantly. In 1960, the cost of 2,4-D at the factory was 40 cents a pound for acid base and 43 cents for ester base. In 1969, the price had fallen to 33.2 cents for acid and 35.5 cents for ester. The chief problem with 2,4-D was its astonishing effectiveness. Because it killed broad-leaved plants, the military used it widely as a defoliant in Vietnam. Opponents of the war attacked both the military and the herbicide.

Research continued, largely with public funds, and increased long before any real public outcry against herbicides. Public funds directly used to carry on weed research in the United States increased: $804,283 in 1950, $1,633,400 in 1955, and up to $4,583,100 in 1962.* Many attacks on weeds and attempts to find new, more selective, and safer herbicides were under way.

The farmers were willing to give up the technological advances of the aircraft industry, the urbanites would surrender herbicides, the bankers would give up insecticides, and the farm equipment industry would pass up atomic power plants. But no one was quite ready to give up his own interests as he saw them, and all saw them differently. Very few seemed seriously interested in undoing what had been done.

By one definition, progress takes place when what is is better than what was. The list of changes has to be totaled up in terms of human welfare, and some judgment has to be made to decide if what has happened is better on balance. The abolition of hunger is now possible, but is the threatened poisoning of the environment worth it? Views vary and, indeed, the facts seem ambiguous, conflicting, or even unknown.

* *Weeds*, 6 (Oct. 1958), 366; *Agricultural Science Review*, 1 (Washington, Fall 1963), 41.

Another view of progress sees it simply as a set of changes unlikely to be undone. The movement from horses to tractors may or may not have been an improvement; however, farmers would be highly unlikely to return to workhorses willingly. Therefore the change from horses to tractors could be considered progress. The historical record has not yet displayed any certainties concerning the biochemical developments of the years from 1945 to 1972.

CHAPTER TWENTY-NINE

ONE WORLD ✻ 1607–1972

BUILDING on the success of its farmers, the United States after nearly 400 years emerged simultaneously as the world's leading industrial nation and one of the most urbanized. Earlier generations might have attributed their success as farmers to God or Providence. Later generations might have explained their success as historical inevitability. Most Americans at any time would have explained the success of U.S. farmers as the natural result of hard work. Free enterprise and freedom of choice, long honored in the United States, seemed an adequate explanation.

In the beginning, Europe had provided the urban markets for colonial American farmers. As American cities grew, farmers selected their enterprises to conform to the changing markets. Technology and science changed farming. In the 17th century, land policy evolved to give farmers land since all efforts to keep the actual farmers from owning their land had failed. Understandably, agriculture and settlement flourished only when farmers themselves owned the land. The European immigrants merged the gardening traditions of the Indians and Africans with their own extensive field husbandry. The native Indian crops of corn and tobacco were adopted for European subsistence and prosperity. Even 17th-century farming brought commercial profits and provided subsistence for other enterprises such as fur trading, lumbering, or fishing. Commercial farming in a fairly primitive technological society encouraged the development of slavery.

American farmers produced surpluses of food that could be marketed abroad. The northern development of grain and livestock commerce in the 18th century also brought changes in tenure and farm size. Farms became larger, and better implements let farmers handle more of the various commodities.

The expanding population and need for more farmland led to an aggressive westward surge, both North and South. The British government sought to regulate commerce (and thus farming) and to halt territorial expansion. The Navigation Acts, amended after 1763, and the Proclamation of 1763 would have made American farming an instrument of British economic growth. The effort infuriated farmers and merchants in the New World. British policy finally resulted in the successful American Revolution.

Success in securing independence led to two kinds of farmer-oriented reform. Land policy evolved to distribute land quickly and cheaply in the American empire, and the United States expanded republican government and full partnership within the empire. Commercial policies, treaties, and tariffs encouraged the export of U.S. farm products. Farmers moved westward, produced more, increased commerce with the cities; and as the 18th century ended, they threatened to seize the port of New Orleans and the territory of Louisiana.

Meanwhile the increasing major European markets and the rise of American cities encouraged ever greater concentration on agricultural commodities such as cotton which suddenly flourished. But advances in transportation, handling, and processing of all commodities also aided the expansion of commercial agriculture. Improved technology and developments in plant husbandry occurred. Westward expansion simultaneously created more surpluses for sale. Substantial additions of farmland as a result of the Louisiana Purchase (1803), the Mexican War (1846–1848), and the Oregon settlement (1846) and more lenient disposal of public lands resulted in more farmers raising more of everything. The process also exacerbated the moral issue of slavery which was linked with commercial farming.

In the 19th century, changes took place in farm technology, in processing, in transportation, and in urbanization, but the anachronism of slavery persisted. The Civil War spurred northern farming as well as industry. The war had disastrous effects on southern farming. The war freed the slaves, but the freedmen rapidly became sharecroppers, saddled with cotton as surely as the landowners were. Meanwhile public land policy was liberalized. The Free Homestead Act was followed by a series of laws to adapt the act to the realities of the unoccupied public domain. The drive to ever greater commercial enterprises in agriculture continued.

As farming became more commercialized, farmers became more dependent on the railroads and other big business. Farmers sought political control of these commercial interests through the Grangers, Greenbackers, Populists, and other groups and parties.

By the mid-19th century, technological progress in farming had become almost self-sustaining, although the application of knowledge depended on the ever-growing, rich markets of America. The endless list of inventions and adoptions includes a host of advances in animal and dairy husbandry. Barbed wire fencing, silos, cream separators, and the gasoline tractor appeared.

Few developments made such technological impact on farming as the tractor. Machinery had to be redesigned for its higher speeds and greater power. Draft animals declined in numbers releasing the acreage formerly used for their feed. Meanwhile, the size of farms increased continuously while the number of farmers declined steadily.

The 20th century introduced total war on an unprecedented scale. Nations at war found that they had to mobilize all resources, including people, to survive. This level of mobilization inevitably affected the technological, scientific, and economic development of American farming. During World War I, the Federal Government experimented with farm price supports. Both world wars brought rationing, wage and price controls, and the allocation of materials and commodities in varying degrees.

Between the two world wars, the United States suffered the worst depression in its history. As a result, the Federal Government became active in all spheres of life. Farmers experienced price supports, subsidies, and crop controls; but much more was involved. New policies on land use and tenure were implemented in the Taylor Grazing Act, the Indian Reorganization Act, and a number of soil conservation acts.

During the world wars and even during the Great Depression, mechanical invention continued with comparatively rapid adoption of field ensilage cutters, automatic hay balers, milking machines, and more powerful tractors. Contour plowing, strip-cropping, and other methods of erosion control were discovered and applied. New fertilizers and vitamin supplements became available to farmers. Hybrid corn; new varieties of cotton, wheat, and other plants; and animal breeds of greater productivity were found or developed. Antibiotics, insecticides, herbicides, and new methods of disease control were discovered. All these developments tremendously increased farm productivity.

Farmers could afford to utilize advances in science and technology. These changes also enabled fewer farmers to supply the nation, and the exodus of farmers to the cities continued. Increasingly the migrants were black and from the South.

Farm prosperity resulted from direct price supports as well as from events of economic importance. Several undeclared wars and relief programs such as the Marshall Plan and Food for Peace provided the most economic help. A resolve to feed the poor in the United States increased the consumers of farm products. The need for food and fiber might have outrun the supply had it not been for increasing productivity of American farmers from 1945 to 1972.

The history of U.S. agriculture showed first an increase of farmers as they settled a vast and nearly empty land. The trend was reversed by the early 20th century. The number of farmers declined as farms increased in size and farm people moved to the cities. Although their numbers fell, the economic and political power of the farmers increased. At some point in the future, farmers may become such a powerful minority that the Federal Government may declare them public servants operating what amounts to a public utility and subject them to the same regulations. By 1972, U.S. agriculture was cartelized, and American farmers were practically part of a public utility already.

TOLSTOY'S THEORY OF HISTORICAL DETERMINISM

Count Leo Tolstoy, musing on what men of the past had not known about themselves, wondered if events in Russia in the late 19th century possibly could have taken another turn. Considering conjunctions of past events, he observed that the more remote the event, the less likely it seemed that it could have been different. To drop Julius Caesar from the history of mankind would so alter everything that all subsequent history would have been quite different. In contrast, Tolstoy noted that some recent event may seem possible to control if the attempt is made. The past seems inevitable, but the present appears subject to human will.

However, according to Tolstoy, control of the present may be an illusion. He thought a multitude of events and even genetic predispositions compel each person to behave in certain ways. In his theory of historical determinism the whole complex of events operating at any moment makes it impossible for anything to happen in any way other than it does. The impelling force is the course of human events. What happens now has to happen because of everything that has gone before. In turn, what happened long ago was predestined because past men and events were under the influence of all that had gone before them.

Therefore, Tolstoy's theory would assert the development of American agriculture was inevitable because of historical determinism. But even if the theory explains reality, filling in the details is not so easy. If historical forces are truly at work, the cause and effect relationships, however complex, can at least be shown in broad outline. The development of American farming was linked to a complex of technological, scientific, political, demographic, and economic changes. The puzzle lies in determining why this complex had such unusual consequences in the United States. Other places, such as Latin America, experienced similar changes without a similar outcome.

The stage of American entry on the world scene may have been decisive. For a nation—as for a person—achievement depends partly on whether the actor arrives on the scene when his talents can be used. Europeans settled in the New World during a crest of technological, demographic, political, and scientific changes in Europe. Americans had all of the benefits and few of the disadvantages of these changes. By the 19th century, American agriculture had developed its own momentum and, with the rest of U.S. industries, kept on booming. In spite of near collapse in the 1930s, the system proved self-regenerating.

The impressive progress of science and technology was mostly of European origin. The emptiness of the continent and its comparatively great distance from Europe enabled the settlers to use fully the advances brought from the Old World. The continuing contact with Europe provided an unceasing exchange of information and goods. The development of large metropolitan centers facilitated taking advantage of the stage of entry. The existence of metropolises in Europe was a precondition of the stage of entry of American farmers, and continued growth of cities had clear reprecussions for them.

Tolstoy, finding the past unchangeable, supposed the present to be as intractable. In 1952, Walt Rostow suggested much the same thing in his theory of self-sustained growth and change in *Process of Economic Growth*. Growth of American agriculture seemed to proceed inexorably.

Predicting the future achievements of science hints at amazing possibilities: "instant" plants that grow to maturity in a month and animals forced to reproduce more often, mature more rapidly, at far less expense. What if pills—vitamins and others—so far replaced food that eating remained only as a pleasant custom? What if atomic-electronic-chemical methods were found to transport irrigation water, and desalination became easy and instantaneous? What if weeds, insects, and diseases could be specifically and fully controlled?

These speculations open up fascinating possibilities for the future whether we subscribe to a theory of historical determinism to explain the development of agriculture in the United States or whether we do not.

BIBLIOGRAPHY ✗

Aー BOUT three-fourths of the titles from my *Bibliography of Books and Pamphlets on the History of Agriculture in the United States, 1607–1967* helped me construct this narrative. The following bibliography contains books mentioned in the text but not appearing in the *Bibliography*. The general topical divisions are: Theory, Directly Agrarian, General Economic and Political, Processing, and Personal Research.

I have benefitted particularly from the research of students and assistants at the Smithsonian and their published reports have been included. The research of my own students has been especially helpful and their articles get special mention.

THEORY

Theory, hypothesis, and interpretations were mostly derived from the following works. Some of them also provided information for the narrative.

Bennett, Merrill K. *The World's Food: A Study of the Interrelations of World Populations, National Diets, and Food Potentials.* New York: Harper, 1954.

Bloch, Marc. *The Historian's Craft.* (Translated by Peter Putnam.) Manchester, England: Manchester University Press, 1954. In the United States, Alfred A. Knopf, Inc.

Collingwood, R. G. *The Idea of History.* Oxford: Clarendon Press, 1951.

Ely, Richard T. and Wehrwein, George S. *Land Economics.* New York: Macmillan, 1940.

Galbraith, John K. *The Affluent Society.* Boston: Houghton Mifflin, 1958.

Kubler, George. *The Shape of Time: Remarks on the History of Things.* New Haven: Yale University Press, 1962.

Mangelsdorf, Paul C. "Review of 'Agricultural Origins and Dispersals,' by Carl O. Sauer," *American Antiquity,* 19 (July, 1953), pp. 87–90.

Peccei, Aurelio. *The Chasm Ahead.* New York: Macmillan, 1969.

Rostow, Walt W. *The Process of Economic Growth.* New York: Norton, 1952.

Sauer, Carl O. *Agricultural Origins and Dispersals.* New York: The American Geographical Society, 1952.

Schlesinger, Arthur M. *Paths to the Present.* New York: Macmillan, 1949.

Servan-Schreiber, Emile. *The American Challenge.* (Translated by Ronald Steel.) New York: Atheneum, 1968.

Toynbee, Arnold J. *A Study of History.* (Abridgement of volumes I to VI by D. C. Somervell.) New York: Oxford University Press, 1947.

Ward, Barbara. *The Rich Nations and the Poor Nations.* New York: Norton, 1962.

Webb, Walter P. *The Great Frontier.* Boston: Houghton Mifflin, 1952.

DIRECTLY AGRARIAN

Sources strictly on the subject of farmers and farming are included here. Most of the information came from these highly specialized narratives and studies.

American Husbandry. Harry J. Carman, editor. New York: Columbia University Press, 1939. Reprint of the edition of 1775.

Animal Diseases: The Yearbook of Agriculture, 1956. Washington: U.S. Department of Agriculture, 1956.

Annual Report of the Secretary of the Treasury, 1868. Washington: 1869.

Ardrey, Robert L. *American Agricultural Implements: A Review of Invention and Development in the Agricultural Implement Industry of the United States.* Chicago: The Author, 1894.

Bailey, Liberty H. (Editor). *Cyclopedia of American Agriculture.* New York: Macmillan, 1907–1909. Vol. I, *Farms* (1907); Vol. II, *Crops* (1907); Vol. III, *Animals* (1908); Vol. IV, *Farm and Community* (1909).

Barger, Harold and Landsberg, Hans H. *American Agriculture, 1899–1939: A Study of Output, Employment and Productivity.* New York: National Bureau of Economic Research, 1942.

Benedict, Murray R. *Farm Policies of the United States: A Study of Their Origins and Development, 1790–1950.* New York: 20th Century Fund, 1953.

Benedict, Murray R. and Stine, Oscar O. *The Agricultural Commodity Programs: Two Decades of Experience.* New York: 20th Century Fund, 1956.

Bidwell, Percy W. and Falconer, John I. *History of Agriculture in the Northern United States, 1620–1860.* Washington: Carnegie Institution of Washington, reprinted, Peter Smith, 1941.

Bogart, Ernest L. *Economic History of American Agriculture.* New York: Longmans, Green, 1923.

Bogue, Allan G. *From Prairie to Farm Belt: Farming on the Illinois and*

Iowa Prairies in the Nineteenth Century. Chicago: University of Chicago Press, 1963.

Bressler, Raymond G., Jr. and Hopkins, John A. *Trends in Size & Production of Aggregate Farm Enterprise, 1909–1936.* Philadelphia: Works Progress Administration, National Research Project, 1938.

Buck, Solon J. *The Agrarian Crusade: A Chronicle of the Farmer in Politics.* New Haven: Yale University Press, 1920.

———. *The Granger Movement: A Study of Agricultural Organization, 1870–1880.* Cambridge: Harvard University Press, 1913.

Carstensen, Vernon R. (Editor). *The Public Lands: Studies in the History of the Public Domain.* Madison: University of Wisconsin Press, 1962.

Church, Lillian M. *History of Corn Planters.* Washington: U.S. Department of Agriculture, Information Series 69, 1935.

———. *Partial History of the Development of Grain Harvesting Equipment.* Washington: U.S. Department of Agriculture, Information Series 72, 1939.

———. *Partial History of the Development of Grain Threshing Implements and Machines.* Washington: U.S. Department of Agriculture, Information Series 73, 1939.

Church, Lillian M. and Tolley, H. R. *Manufacture and Sale of Farm Equipment in 1920.* Washington: U.S. Department of Agriculture, Circular 212, 1922.

Clawson, Marion. *Uncle Sam's Acres.* New York: Dodd, Mead, 1951.

———. *The Western Range Livestock Industry.* New York: McGraw-Hill, 1950.

Cooper, Martin R., Barton, Glen T., and Brodell, Albert P. *Progress of Farm Mechanization.* Washington: U.S. Department of Agriculture, Miscellaneous Publication 630, 1947.

Dana, Samuel T. *Forest and Range Policy: Its Development in the United States.* New York: McGraw-Hill, 1956.

East, Edward and Jones, Donald F. *Inbreeding and Outbreeding.* Philadelphia: Lippincott, 1919.

Elwood, Robert B., Arnold, L. E., Schmutz, D. C., and McKibben, E. G. *Changes in Technology and Labor Requirements in Crop Production: Wheat and Oats.* Philadelphia: Works Progress Administration, National Research Project, 1939.

Elwood, Robert B., Knowlton, H. E., and McKibben, E. G. *Changes in Technology and Labor Requirements in Crop Production: Potatoes.* Philadelphia: Works Progress Administration, National Research Project, 1938.

Elwood, Robert B., Lewis, Arthur A., and Strubble, Ronald A. *Changes in Technology and Labor Requirements in Livestock Production: Dairying.* Washington: Works Progress Administration, Report, 1941.

Farmers in a Changing World: Yearbook of Agriculture, 1940. Washington: U.S. Department of Agriculture, 1940.

Gittins, Bert S. *Land of Plenty.* Chicago: Farm Equipment Institute, 1950.

Grass: The Yearbook of Agriculture, 1948. Washington: U.S. Department of Agriculture, 1948.

Gray, Lewis C. *History of Agriculture in the Southern United States to 1860.* Washington: Carnegie Institution of Washington, 1933. 2 volumes.

Gray, Roy B. *Development of the Agricultural Tractor in the United States.* Washington: U.S. Department of Agriculture, 1954. 2 volumes.

Harding, Thomas Swann. *Two Blades of Grass: A History of Scientific Development in the United States Department of Agriculture.* Norman: University of Oklahoma Press, 1947.

Hargreaves, Mary W. *Dry Farming in the Northern Great Plains, 1900–1925.* Cambridge: Harvard University Press, 1957.

Haystead, Ladd, and Fite, Gilbert C. *Agricultural Regions of the United States.* Norman: University of Oklahoma Press, 1955.

Hibbard, Benjamin H. *A History of the Public Land Policies.* New York: Macmillan, 1924.

An Historical Survey of American Agriculture. Washington: U.S. Department of Agriculture, Yearbook Separate 1783, 1941.

Insects: Yearbook of Agriculture, 1952. Washington: U.S. Department of Agriculture, 1952.

McGovern, George S. *War Against Want: America's Food for Peace Program.* New York: Walker and Co., 1964.

McKibben, Eugene G. and Griffin, R. Austin. *Changes in Farm Power and Equipment: Tractors, Trucks, and Automobiles.* Philadelphia: Works Progress Administration, National Research Project, 1938.

McKibben, Eugene G., Hopkins, John A., and Griffin, R. Austin. *Changes in Farm Power and Equipment: Field Implements.* Philadelphia: Works Progress Administration, National Research Project, 1939.

Mangelsdorf, Paul C. *Plants and Human Affairs.* Notre Dame: University of Notre Dame Press, 1952.

Nourse, Edwin G. *American Agriculture and the European Market.* New York: McGraw-Hill, 1924.

Peffer, E. Louise. *The Closing of the Public Domain: Disposal and Reservation Policies, 1900–1950.* Stanford: Stanford University Press, 1951.

Peterson, Gale E. "The Discovery and Development of 2,4-D," *Agricultural History,* 41 (July, 1967), 243–53.

Phillips, Ulrich B. *Life and Labor in the Old South.* Boston: Little, Brown, 1929.

Rasmussen, Wayne D. *Readings in the History of American Agriculture.* Urbana: University of Illinois Press, 1960.

Robbins, Roy M. *Our Landed Heritage: The Public Domain, 1776–1936.* Lincoln: University of Nebraska Press, 1962. Reprint of the Princeton University Press edition of 1942.

Rogin, Leo. *The Introduction of Farm Machinery in its Relation to the Productivity of Labor in Agriculture of the United States During the Nineteenth Century.* Berkeley: University of California Press, 1931.

Schmidt, Louis B. and Ross, Earle D. *Readings in the Economic History of American Agriculture.* New York: Macmillan, 1925.

Shannon, Fred A. *The Farmer's Last Frontier, Agriculture, 1860–1897.* New York: Farrar & Rinehart, 1945.

Sharrer, G. Terry. "The Indigo Bonanza in South Carolina, 1740–90," *Technology and Culture,* 12 (July, 1971), 447–55.

Shaw, Eldon E. and Hopkins, John A. *Trends in Employment in Agriculture,*

1909–36. Philadelphia: Works Progress Administration, National Research Project, 1938.

Street, James H. *New Revolution in the Cotton Economy: Mechanization and Its Consequences.* Chapel Hill: University of North Carolina Press, 1957.

Summons, Terry G. "Animal Feed Additives, 1940–1966," *Agricultural History,* 42(October, 1968), 305–13.

Towne, Charles W. and Wentworth, Edward N. *Pigs: From Cave to Corn Belt.* Norman: University of Oklahoma Press, 1950.

U.S. Department of Agriculture. *Agricultural Statistics.* Washington: 1936 to 1972. (Annual.)

———. *Changes in Farm Production and Efficiency: A Summary Report, 1966.* Washington: U.S. Department of Agriculture, Statistical Bulletin 233, 1966.

———. *Technology on the Farm.* Washington: 1940.

U.S. Patent Office. *Annual Report of the Commissioners of Patents.* Washington: Patent Office, 1838–1862.

Wertenbaker, Thomas J. *The Planters of Colonial Virginia.* Princeton: Princeton University Press, 1922.

Wessel, Thomas R. "Prologue to the Shelterbelt, 1870–1934," *Journal of the West,* 6 (January, 1967), 119–34.

———. "Roosevelt and the Great Plains Shelterbelt," *Great Plains Journal,* 8 (Spring, 1969), 55–74.

GENERAL ECONOMIC AND POLITICAL

Economic and political information about the society and about the farmers was obtained mostly from the following books.

Allen, Frederick L. *Since Yesterday: The Nineteen-thirties in America, September 3, 1929–September 3, 1939.* New York: Harper, 1940.

Billington, Ray A. *Westward Expansion: A History of the American Frontier.* New York: Macmillan, 1960.

Bining, Arthur C. and Cochran, Thomas C. *The Rise of American Economic Life.* New York: Scribner's Sons, 1964.

Blake, Nelson M. *A Short History of American Life.* New York: McGraw-Hill, 1952.

Carson, Rachel L. *Silent Spring.* Boston: Houghton Mifflin, 1962.

Channing, Edward. *A History of the United States.* Vol. II, *A Century of Colonial History, 1660–1760.* New York: Macmillan, 1948.

Clark, Thomas D. *Frontier America.* New York: Scribner's Sons, 1959.

Craf, John R. *Economic Development of the United States.* New York: McGraw-Hill, 1952.

Faulkner, Harold U. *American Economic History.* New York: Harper, 1960.

Fite, Gilbert C. and Reese, Jim E. *An Economic History of the United States.* Boston: Houghton Mifflin, 1965.

Genung, A. B. *Food Policies During World War II*. Ithaca: Northeast Farm Foundation, 1951.

Goldman, Eric F. *The Crucial Decade: America, 1945–1955*. New York: Knopf, 1956.

Handlin, Oscar. *The Uprooted: The Epic Story of the Great Migrations That Made the American People*. Boston: Little, Brown & Co., 1951.

Hicks, John D. *The American Nation: A History of the United States from 1865 to the Present*. Cambridge: Houghton Mifflin, 1955.

———. *The Federal Union: A History of the United States to 1865*. Cambridge: Houghton Mifflin, 1952.

Kemmerer, Donald L. and Jones, C. Clyde. *American Economic History*. New York: McGraw-Hill, 1959.

Kirkland, Edward C. *A History of American Economic Life*. New York: Appleton-Century-Crofts, 1951.

Kolko, Gabriel. *Wealth and Power in America: An Analysis of Social Class and Income Distribution*. New York: Frederick A. Praeger, 1962.

Leuchtenburg, William E. *Franklin D. Roosevelt and the New Deal, 1932–1940*. New York: Harper & Row, 1965.

McGrane, Reginald C. *The Economic Development of the American Nation*. Boston: Ginn, 1950.

Merk, Frederick. *Manifest Destiny and Mission in American History*. New York: Knopf, 1963.

Morison, Samuel E. and Commager, Henry S. *The Growth of the American Republic*. New York: Oxford University Press, 1954. 2 volumes.

Ross, Earle D. *Democracy's College: The Land-Grant Movement in the Formative Stage*. Ames: Iowa State College Press, 1942.

Shannon, Fred A. *America's Economic Growth*. New York: Macmillan, 1951.

Taylor, George R. *The Transportation Revolution, 1815–1860*. New York: Rinehart, 1951.

Taylor, George R. and Neu, Irene D. *The American Railroad Network, 1861–1890*. Cambridge: Harvard University Press, 1956.

U.S. Bureau of Land Management. *Public Land Statistics*. Washington: U.S. Department of the Interior, 1962.

U.S. Bureau of Public Roads. For manuscript figures for roads, sizes, and area taken in roads. Washington: U.S. Department of Transportation.

U.S. Bureau of the Census. *Historical Statistics of the United States: Colonial Times to 1957*. Washington: U.S. Department of Commerce, 1961.

———. *Historical Statistics of the United States, 1789–1945*. Washington: U.S. Department of Commerce, 1949.

———. *Statistical Abstract of the United States*. Washington: 1878 to 1972. Annual. From 1878 to 1902 published by the Treasury Department; from 1903 to 1911, the Department of Commerce and Labor; and from 1912 to 1972, the Department of Commerce.

United States Census. Printed reports, 1850 to 1969.

U.S. Foreign Agricultural Service. *Foreign Agricultural Trade: Statistical Handbook*. Washington: U.S. Department of Agriculture, Statistical Bulletin 179, 1956.

Wecter, Dixon. *The Age of the Great Depression, 1929–1941*. New York: Macmillan, 1948.

Wertenbaker, Thomas J. *The Old South: The Founding of American Civilization.* New York: Scribners, 1942.

PROCESSING AND HANDLING

The handling of farm products, whether for ease of transport or of consumption, comes under the rubric of processing. These books provided most of the information on processing.

Anderson, Oscar. *Refrigeration in America: A History of a New Technology and Its Impact.* Princeton: Published for the University of Cincinnati by Princeton University Press, 1953.

Corey, Lewis. *Meat and Man: A Study of Monopoly, Unionism, and Food Policy.* New York: Viking Press, 1950.

Cummings, Richard O. *The American and His Food.* Chicago: University of Chicago Press, 1940.

de Kruif, Paul H. *Hunger Fighters.* New York: Harcourt-Brace, 1928.

Kohls, Richard L. *Marketing of Agricultural Products.* New York: Macmillan, 1955.

Pirtle, Thomas R. *History of the Dairy Industry.* Chicago: Mojonnier Brothers Co., 1926.

Power to Produce: Yearbook of Agriculture, 1960. Washington: U.S. Department of Agriculture, 1960.

Purcell, Margaret. *Statistical Findings of Survey of Transportation from Farms to Initial Markets.* Washington: U.S. Department of Agriculture, Bureau of Agricultural Economics, 1949.

Reall, Joseph H. *Dairying and Dairy Improvements.* New York: 1882.

Storck, John and Teague, Walter D. *Flour for Man's Bread.* Minneapolis: University of Minnesota Press, 1952.

Tressler, Donald K. and Evers, Clifford F. *The Freezing Preservation of Foods.* Westport, Connecticut: Avi Publishing Co., 1957.

U.S. Tariff Commission. *Synthetic Organic Chemicals: United States Production and Sales.* Washington: 1952.

Wagner, Arthur F. and Folkers, Karl. *Vitamins and Coenzymes.* New York: Wiley, 1966.

Yearbook of the United States Department of Agriculture, 1899. Washington: 1900.

PERSONAL RESEARCH

A major influence in theory and in information has been my own research and writing. The relevant results of that work are:

Schlebecker, John T. *Agricultural Implements and Machines in the Collection of the National Museum of History and Technology.* Washington: Smithsonian Institution Press, 1972.

———. "Agriculture in Western Nebraska, 1906–1966," *Nebraska History,* 48 (Autumn, 1967), 249–66.

———. *Bibliography of Books and Pamphlets on the History of Agriculture in the United States, 1607–1967.* Santa Barbara: Clio Press, 1969.

———. *Cattle Raising on the Plains, 1900–1961.* Lincoln: University of Nebraska Press, 1963.

———. "The Combine Made in Stockton," *The Pacific Historian,* 10 (Autumn, 1966), 14–21.

———. "Dairy Journalism: Studies in Successful Farm Journalism," *Agricultural History,* 31 (October, 1957), 23–33.

———. "Farmers in the Lower Shenandoah Valley, 1850," *The Virginia Magazine of History and Biography,* 79 (October, 1971), 462–76.

———. "The Federal Government and Cattlemen on the Plains, 1900 to 1945," *Probing the American West.* Santa Fe: Museum of New Mexico Press, 1962.

———. "Grasshoppers in American Agricultural History," *Agricultural History,* 27 (July, 1953), 85–93.

———. "The Great Holding Action: The NFO in September, 1962," *Agricultural History,* 39 (October, 1965), 204–13.

———. *A History of American Dairying.* Chicago: Rand McNally, 1967.

———. *A History of Dairy Journalism in the United States, 1810–1952.* With Andrew W. Hopkins. Madison: University of Wisconsin Press, 1957.

———. "Pliant Prairie: One Plant's Influence on One Prairie State," *Montana, The Magazine of Western History,* 8 (January, 1958), 30–41.

———. "The World Metropolis and the History of American Agriculture," *Journal of Economic History,* 20 (June, 1960), 187–208.

INDEX ❧